# PSYCHEDELICS

**Also by Professor Nutt**

*Drink?*
*Drugs Without the Hot Air*
*Nutt Uncut*
*Cannabis (Seeing Through the Smoke)*
*Brain and Mind Made Simple*

# PSYCHEDELICS

## The Revolutionary Drugs That Could Change Your Life— A Guide from the Expert

## PROFESSOR DAVID NUTT

hachette
BOOKS

New York

Copyright © 2024 by David Nutt
Coauthor: Brigid Moss
Figures by Cali Mackrill at Malik & Mack
First published in Great Britain in 2023 by Yellow Kite
An imprint of Hodder & Stoughton
An Hachette UK company

Cover Photographs: © Davide Illini / Stocksy United, Sean Sinclair / Unsplash
Cover copyright © 2024 by Hachette Book Group, Inc.

Hachette Go, an imprint of Hachette Books
Hachette Book Group
1290 Avenue of the Americas
New York, NY 10104
HachetteGo.com
Facebook.com/HachetteGo
Instagram.com/HachetteGo

First Edition: January 2024

Published by Hachette Go, an imprint of Hachette Book Group, Inc. The Hachette Go name and logo is a trademark of the Hachette Book Group.

The Hachette Speakers Bureau provides a wide range of authors for speaking events. To find out more, go to hachettespeakersbureau.com or email HachetteSpeakers@hbgusa.com.

Hachette Go books may be purchased in bulk for business, educational, or promotional use. For information, please contact your local bookseller or Hachette Book Group Special Markets Department at special.markets@hbgusa.com.

The publisher is not responsible for websites (or their content) that are not owned by the publisher.

Library of Congress Cataloging-in-Publication Data has been applied for.

ISBNs: 9780306835285 (trade paperback); 9780306835308 (ebook)

Library of Congress Control Number: 2023946958

Printed in the United States of America

LSC-C

Printing 1, 2023

# CONTENTS

# INTRODUCTION

Psychedelics are the new revolution in neuroscience and psychiatry. In the past few years, the world of psychedelics has changed dramatically. For 50 years, the global War on Drugs, started in the U.S. by President Nixon in the 1960s, outlawed these compounds in the most draconian fashion.

Now, President Biden has stated that psilocybin and MDMA will be medicines in the U.S. within a couple of years. Hopefully, in a few years Europe will follow. Already, magic mushrooms are now legal in certain towns and states in the U.S. and Canada. As I write, Australia has, in February 2023, revised their Therapeutic Goods Act so that psilocybin and MDMA are medicines from July 1, 2023. Psilocybin as magic mushrooms will shortly become available in state-ratified clinics in Oregon and will likely soon be in California and British Columbia too.

This new take on psychopharmacology has consumed the past 15 years of my research career and is my most significant body of work. In 2018, along with Dr. (now Professor) Robin Carhart-Harris, I set up the first academic psychedelic research center in the world, part of a global revival of research into psychedelics including ketamine, LSD, magic mushrooms and MDMA. What used to be a category of drugs that were considered dangerous and addictive, so that even researching them could derail a scientist's career, has exploded into the revolution in neuroscience and psychiatry that it is today.

In 2006 our group, working with the Beckley Foundation, set out to discover more about brain function underpinning the nature of the psychedelic experience. At the time, I had no inkling that I would soon be writing grants to study psychedelics as a treatment for depression. Our research produced the biggest surprise of my career. Sixties psychedelic figurehead Dr. Timothy Leary postulated that people should "turn on, tune in, drop out," but to our amazement, psychedelics did not turn you on at all. Or more precisely, they didn't turn on the brain. Rather they turned off parts of the brain, in particular the parts driving depressive thinking. This insight led to us doing studies in people with treatment-resistant depression, and these revealed that a single psilocybin trip was the most powerful single treatment ever for this common and disabling condition.

Psychedelics are looking promising for other conditions too, such as PTSD, OCD, eating disorders and some pain conditions. They offer hope to the large proportion of patients for whom current treatments have failed or are inadequate, and to psychiatrists and psychologists too. Because these new treatments are the first significant breakthroughs in psychiatric treatments in the past 50 years, despite major advances in neuroscience and brain imaging. Already ketamine, which has a psychedelic-like effect, is being used for treatment-resistant depression. This is possible because ketamine, having been used as an anesthetic for decades in many countries, is a licensed medicine, unlike psychedelics and MDMA.

Our group has become one of the world's leading centers for psychedelic research, in particular by pioneering the use of brain imaging to understand their therapeutic mechanisms as well as their effects on consciousness. Because as well as being potential medicines, psychedelics are tools that teach us about brain mechanisms and consciousness. As William James said over a century ago, after his psychedelic experiences:

> Our normal waking consciousness...is but one special type of consciousness, whilst all about it, parted from it by the filmiest of screens, there lie potential forms of consciousness entirely different...No account of the universe in its totality can be final which leaves these other forms of consciousness quite disregarded. How to regard them is the question—for they are so discontinuous with ordinary consciousness.

I agree with James—these altered states are scientific questions as important as the study of subatomic particles or the origins of the universe. Modern neuroimaging techniques allow us to look at the effects of psychedelics on different brain regions and, more importantly, how connectivity between these regions is altered during the trip and even afterward.

You could say these techniques are to psychedelic consciousness research what the Large Hadron Collider is to particle physics. And like with particle physics, I suspect we are only at the beginning of our understanding. J. B. S. Haldane, the great British medical scientist of the early twentieth century, said: "Now, my own suspicion is that the universe is not only queerer than we suppose, but queerer than we *can* suppose." Perhaps the same is true for the brain.

# Chapter 1

# WHY PSYCHEDELICS ARE BACK

**WE ARE ON** the cusp of a revolution in psychiatric medicine. After 50 years of prohibition, criminalization and stigmatization, science is finally showing that psychedelics are not usually dangerous or harmful. Instead, when used according to tested, safe and ethical guidelines, they are the next revolution in mental health treatment.

You might have picked up this book because you've heard of the groundbreaking research into psilocybin for treatment-resistant depression. Or how MDMA is being used as therapy for post-traumatic stress disorder (PTSD). From July 1, 2023, both will be medicines in Australia,[1] and MDMA soon will be in the U.S. and Canada too.

Or perhaps you picked it up because a friend has told you their mental health has been transformed by going on an ayahuasca retreat in the Amazon region. Or perhaps that microdosing LSD or magic mushrooms made them more productive or creative or improved their mood. You might be wondering if taking psychedelics really is safe and if they work. Or even be considering if it's worth traveling to territories where people are allowed to take some psychedelics, such as the Netherlands, Jamaica, Oregon or Colorado, in order to access their benefits legally.

This book is the fruit of 15 years of research and scientific innovation. It will answer your questions about psychedelics and psychedelic-like drugs, including magic mushrooms (psilocybin), ayahuasca and LSD. These drugs are now firmly in the mainstream of research into the treatment of depression, addiction and PTSD and are looking promising for other conditions too, including OCD, eating disorders and some pain conditions.

But it's not just about the drugs. The best results are achieved by combining psychedelics and psychotherapy, a completely new approach to mental illness that gets the best out of both treatments. As well as being a neuroscientist and psychopharmacologist (someone who investigates drugs and the brain), I'm also a psychiatrist. So I know how important it is that these drugs bring the hope of an effective treatment to the large proportion of patients for whom current treatments have failed or are inadequate.

The book will describe what we—my team at Imperial College and I—have discovered about how psychedelics work on the brain (see Chapter 4), and how we found that they work very differently from the antidepressants that are currently the mainstay of clinical psychiatry (see Chapter 5). This is important because the more approaches medicine has to treat a disorder, the better the potential outcome for the patient. Studies are showing that psychedelics work fast and usually just require a couple of doses, whereas the older drugs take weeks if not months to work. You can rapidly get people well, which conventional treatments in psychiatry rarely do.

## THE RETURN OF THE TRIP

As well as in medicine, this class of drugs, despite having been illegal in most countries for 50 years, is now showing up in culture, academia and business. It started with Michael Pollan's excellent book *How to Change Your Mind*, which led to his Netflix show. Now, there's Prince Harry writing openly about his experiences with ayahuasca and magic mushrooms in his book *Spare*. But more importantly for the near future of psychedelics as medicines, the establishment is getting on board too. Professor Paul Summergrad, a previous president of the American Psychiatric Association, has described how early experiences with LSD led him to becoming a psychiatrist:

"It was a deeply mystical experience, and it also changed my thinking about the self. It made me think a lot about neurobiology and consciousness, because if a tiny dose of a drug like this could change one's perception so profoundly, what did that mean regarding how we understand the mind–brain relationship and, relevant to psychiatry, the etiology of mental illness?"[2]

The research renaissance was led by a wave of new university research groups, including mine, followed by a second wave, this one of specialist pharmaceutical companies. Several of the companies dedicated to this research are now valued at billions of dollars. This acceptance by the mainstream might be because the generation now in power are likely to have experimented with psychedelics in their youth. One of the first to talk about this publicly was Steve Jobs, who said taking LSD was one of the most important experiences of his life. "I have no words to explain the effect the LSD had on me, although I can say it was a positive life-changing experience for me and I am glad I went through that experience."[3]

The story of how psychedelics came to prominence in the 1960s is a compelling one. In 1943, the psychedelic effects of LSD were discovered by accident when the scientist who synthesized it, Albert Hofmann, accidentally ingested some after touching it with his fingertips.[4]

Later, Hofmann also identified the similar active ingredient of magic mushrooms as psilocybin. It seems incredible now, but during the 1950s Sandoz marketed these compounds as Delysid and Indocybin respectively, sending them out to researchers around the world in an attempt to find out how they could be used in medicine.

This led to the first psychedelic research boom of the 1950s and 1960s, hundreds of clinical trials and thousands of case reports of psychedelic drugs, in particular of LSD.[5] It's estimated that during this period the National Institutes of Health, the research funding arm of the U.S. government, gave over 130 grants.[6]

Psychedelics were a step change in the development of biological psychiatry. By showing that a chemical can perturb the mind and brain, they showed it was possible to treat disorders of the mind with drugs.

Modern psychiatric research stands on the shoulders of this early work. Psychiatrists began to use psychedelics to allow patients to access repressed memories and emotions, to unblock people who weren't moving forward in psychotherapy. It was around this time that the name psychedelics was coined, meaning "mind-manifesting." As you'll read in Chapter 7, one of the most successful applications was in treating addiction to alcohol.

There was a dark side to mid-century psychedelic research too. In the U.S., a large part of it was carried out or sponsored by the CIA, the army and other government agencies. They tested LSD on both soldiers and civilians, both with and without consent (see Chapter 13), as a potential weapon in the Cold War. There were stories that the Soviets were going to use LSD as a chemical weapon, to spray Western soldiers with it to disable them, or put it in the water supply to knock out entire cities of people.[7,8] These kinds of fears drove research in the West—the UK army did their own testing too. Although a lot of this research was unethical, even by the standards of the time, it did mark the beginning of the examination of psychedelics' effects on the brain.

When LSD escaped from the lab in the 1960s, it fueled an explosion of creativity in art and music and of reform in politics. In the U.S. the Haight-Ashbury district in San Francisco became the center of the new youth counterculture movement, with the young people "turning on" and dropping out of normal life. There were demonstrations against the Vietnam War, environmental marches, civil rights protests and the birth of feminism, or "women's lib" as it was then called.

I was a little too young to be a part of this—I got my O level results in the summer of 1967, aka the Summer of Love. I did make a ball-and-stick model of LSD for parents' day at school, a small rebellion (I had wanted to make the substance itself in my A-level chemistry class). I do remember that, at that time, you couldn't escape the feeling that the world was changing for the better. You don't have to take LSD to know that "make peace, not war" is a good thing.

There were the first music festivals, starting with the Human Be-In in San Francisco. Music was a key part of many people's first experiences of psychedelics (Chapter 6 describes why it's important in psychedelic-assisted therapy too). The drugs birthed the Grateful Dead, the Doors, Jimi Hendrix and the Beatles' *Sgt. Pepper's Lonely Hearts Club Band*. The magic acid test bus began touring, spreading its radical message of peace, love and music.

Paul McCartney later confessed that he was an atheist until he took dimethyltryptamine (DMT), a short-acting psychedelic with effects similar to LSD (see page 31). He said that he was "immediately nailed to the sofa—and I saw God, this amazingly huge towering thing, and I was humbled." He described his vision as: "huge...a massive wall that I couldn't see the top of, and I was at the bottom."[9] (More on the spiritual side of psychedelics in Chapter 9.)

Then, in 1967, LSD was banned in the U.S. as a Schedule 1 drug, branded highly dangerous, addictive and without any special medical use. The U.S. authorities led a global campaign to ban psychedelics and within a few years it was classified under Schedule 1 of the 1971 United Nations convention on drugs too. Most countries followed the UN's lead. Other psychedelics, particularly psilocybin and mescaline, were also included in the ban, even though there was very little evidence of their use, let alone harm.

History has shown us that psychedelics were not banned because they were harmful. They were banned because they were changing the way people thought about the big issues of the world, which made them terrifying for the establishment.

Psychedelics have an ability to change unhelpful thoughts that has turned out to be very useful in treating psychological problems including depression and addiction (see Chapters 5 and 7). But the establishment didn't see this quality as useful when it was linked to social unrest challenging the existing government, especially in relation to military foreign policy and the dominance of capitalism. The ban was an attempt by the U.S. government to regain control over this social unrest; they couldn't ban anti–Vietnam War protests, so they banned LSD.

These drugs clearly should not have been put into Schedule 1; they were not that dangerous, and they did have medical uses. Eventually the protests did stop the Vietnam War, but the ban on psychedelics has stayed—until now.

The scheduling effectively censored research on psychedelics for over 50 years. The number of scientific studies dropped to virtually zero. And the existing research, now recast as both illegal and dangerous, was ignored and disappeared from the canon of acceptable psychiatric research.

When I was training as a psychiatrist in the 1970s, I didn't hear about any of the positive research findings. In order to justify the psychedelics ban, governments had sponsored research to show their risk of causing harm, and so all the focus was on this. Media-spread scare stories did a good job of propagating this information, for example, warning that if you took LSD you would look into the sun until you went blind, or think you could fly and jump out a window. Articles warned of LSD damaging chromosomes and corrupting youth. And most of all, they warned of people not being able to get back to normal after tripping or being driven mad (for more, see Chapter 13).

Figure 1: The impact of scheduling on psychedelic publications, from 1950 to 2016

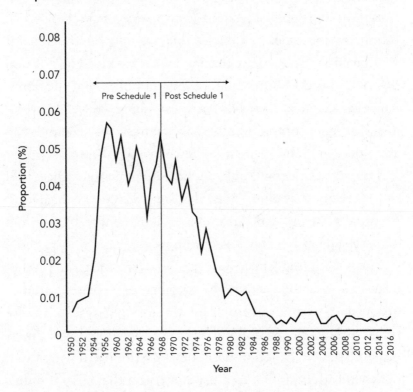

Source: Adapted from Rucker JJH, Iliff J, Nutt DJ, 2018, Psychiatry & the Psychedelic Drugs. Past, Present & Future, *Neuropharmacology* 142: 200–218, ISSN: 0028-3908.

# RESURRECTING PSYCHEDELICS THROUGH RESEARCH

Because psychedelics' bad reputation persists, this book will also dispel some myths and misperceptions (see Chapters 12 and 13). It's true that they are powerful drugs and need to be treated seriously, but most people still believe the claims that these drugs are very dangerous, addictive and without medical value. I spend a lot of my time educating so-called experts who really should know better, including many senior psychopharmacologists and psychiatrists. Then there are the politicians and bureaucrats who believe that the UK can't make these drugs into medicines, because it would go against the UN. This isn't true. On top of this, there's still a significant section of the media that is happy to publish sensationalist scare stories about drugs.

It was during my work as a government adviser, starting in the 1990s, that the very wide gulf between the actual harm of psychedelics and their illegal status became clear to me.

For example, in the UK probably about a million people each year were picking and using magic mushrooms during the autumn growing season. But in 2005, fresh magic mushrooms were put into Class A (dried were already in there), alongside heroin and cocaine, the class with the highest penalties for possession.

This assessment of their safety was clearly not logical. Evidence suggests that for as long as human beings have been consuming plants, they've been using them to alter consciousness, for pleasure and escape but also, more profoundly, to explore what their brain can do, to mark a coming of age, for cohesion of the group and for spiritual and religious reasons.

Looking back to prehistory, there's a theory that the origins of the Hindu religion come from the use of various mind-altering drugs collectively known as soma (a name later used by Aldous Huxley in his novel *Brave New World* for his people-pacifying drug). The exact composition of soma is unknown as its few descriptions are in very old texts; in fact it's still being debated. However, the psychedelic mushroom expert R. Gordon Wasson believed that it probably included extracts of psychedelic mushrooms such as psilocybin-containing magic mushrooms, or *amanita muscaria* (fly agaric, see Chapter 3), as well as other plants such as cannabis.[10]

At a similar time, around 1500 BC, the ancient Greeks were thought to be using psychedelics in a secret religious festival in honor of the goddess Demeter and her daughter Persephone. The Eleusinian Mysteries were held yearly for 2,000 years, from around 1600 BC. During the ten-day festival, people would fast while walking the eleven miles from Athens to the Temple of Demeter in Eleusis, then take a sacred psychedelic drink, *kykeon*, thought to be wine with the addition of the ergot fungus from cereal crops, which makes molecules similar to LSD.[11]

The first modern historical evidence came from the conquistadors, the Spanish who invaded South America. They reported the use of psychedelic drugs, including magic mushrooms, ayahuasca and mescaline, in various parts of South and Central America. The Spanish attempted to wipe out all records and knowledge of psychoactive plants, making their use punishable by death,[12] but these practices adapted to and survived the attempt. Indeed, as the Spanish imposed their language and culture on the existing population, rather than stopping using mescaline, the indigenous peoples, amusingly, renamed the cactus San Pedro after St. Peter, who opens the gates of heaven. More recently, the popularity of ancient ayahuasca religious ceremonies has gone global.

There's a parallel with my government work for the Advisory Committee on the Misuse of Drugs (ACMD) in the 2000s. Then, the drug that the government was trying and failing to stop people using was MDMA. Our official research showed that ecstasy, despite very sadly causing some deaths in young people, was not as dangerous as its classification suggested. It was in Class A, carrying the most serious penalties. This motivated me to start my work establishing fair and logical drug laws based on scientific evidence of harm.

As you'll read in Chapter 12, I was sacked by the government in 2009 for speaking out about the relative harm of different drugs, in particular for saying that the harms of MDMA were being overstated. That led me to produce the definitive way of assessing drug harm.

Three years later, that work made me confident that not only was it safe to research MDMA on humans but the same was true of psilocybin too.

As you'll see in Chapters 6 and 12, the past 15 years of research has produced a systematic and detailed analysis of the claimed harm and benefits of psychedelics. It has shown that they are not dangerous when used properly as medicines, and that their risks are not barriers to them being made medicines as long as they are used with care. In truth, all medicines are dangerous if used improperly.

## INTO THE BRAIN

I started researching psychedelics because I wanted to know more about the brain and consciousness. To quote the psychedelics chemist Alexander (Sasha) Shulgin aka Dr. Ecstasy, "I've always been interested in the machinery of the mental process."[13]

Psychedelics change the brain in the most profound and interesting way of any drug class. The most eloquent and compelling insights into these profound effects are to be found in the revolutionary 1954 book by Aldous Huxley, *The Doors of Perception.*

Huxley's ideas are so interesting because they are so prescient, foreshadowing later scientific work, even though he was writing before very much was known about the brain.

> To be shaken out of the ruts of ordinary perception, to be shown for a few timeless hours the outer and the inner world, not as they appear to an animal obsessed with survival or to a human being obsessed with words and notions, but as they are apprehended, directly and unconditionally, by Mind at Large—this is an experience of inestimable value to everyone and especially to the intellectual.[14]

In the 1950s, Albert Hofmann deduced that psilocybin, the active ingredient in magic mushrooms, has a chemical structure similar to the neurotransmitter serotonin, and that a part of LSD looks like it too.[15] But at the time, it wasn't known what serotonin did or how it acted in the brain.

The 1970s brought a sea change in brain science: the discovery that the brain is largely a chemical machine. That the information that creates our thoughts, functions and processes travels along our neurons via electricity but jumps the gap—the synapse—between neurons using neurotransmitters, a group of chemicals including serotonin.

Later a new technique called *gene cloning* led to the discovery that serotonin has 15 different kinds of receptors in the brain, and that psychedelics produce their typical effects by binding to one of them, the 5-HT2A receptor (usually called the 2A receptor).

Because they perturb the mind, psychedelics are an important tool in understanding the role of some serotonin receptors in human consciousness and also, crucially, in mood, as well as serotonin's overall contribution to the working of the human mind.

We now have neuroimaging techniques that can reveal what happens when a psychedelic slots into the receptors. We can use magnetoencephalography (MEG) to look at electrical brain waves.

Functional Magnetic Resonance Imaging (fMRI) shows us how the different regions of the brain are changed by psychedelics, and how the connections between them are altered. Positron Emission Tomography (PET) shows where the drugs go in the brain, which parts of the brain they bind to, and how long they sit in the brain.

Because psychedelics are still in Schedule 1, there are still multiple barriers to doing this research, not least the cost. The ban, which has lasted 50 years, has thus created a 50-year hole in research. This is the worst censorship of research ever because its impact has been global. The fact that psychedelics are about to become medicines shows that they should never have been banned in the first place—probably not even for recreational use, but certainly not as medicines. There are 50 years' worth of people with mental illnesses and addictions who have been denied the best treatment that could have been available.

This book is part of the reversal of this censorship—the psychedelic renaissance. This is one of the most exciting areas of modern science. By the end, I trust you will understand why this is the case.

# Chapter 2

# WHAT IS A PSYCHEDELIC? MEET THE CLASSICS: LSD, MAGIC MUSHROOMS, AYAHUASCA

**AT FIRST GLANCE,** the members of this family of drugs might appear unrelated to one another. As well as LSD and magic mushrooms, it includes the South American ceremonial tea ayahuasca, the drug of choice of the modern generation of spiritual seekers. Then there's the "god molecule" Toad (which really does come from a toad). Finally, there is mescaline, the first psychedelic to be chemically identified and one of the ingredients in the "savage journey across America" described by Hunter S. Thompson in his novel *Fear and Loathing in Las Vegas*.

Despite their widely varying backstories, histories and reputations, all the classics work in the same way. Their alternative name—the serotonergic psychedelics—explains why. Each classic psychedelic molecule has a part of its structure that mimics serotonin, allowing it to bind to one or more of serotonin's receptors in the brain and the gut. Serotonin (5-hydroxytryptamine or 5-HT) is a neurotransmitter that has roles in many different brain functions, including sleep, memory and learning, mood and emotions, sexual behavior, hunger and perceptions. Serotonin is also the target for other drug treatments in psychiatry, including SSRIs (selective serotonin reuptake inhibitors), which are used in depression and anxiety.

Receptors are the target proteins that a neurotransmitter acts on. The human brain has at least 15 different subtypes of serotonin receptors and most of them do different things. And the classic psychedelics are a group because they all produce their psychedelic effects by binding to the 5-HT2A or 2A receptor.

Most of the classics are currently illegal in most Western countries. Despite the regulatory and financial barriers to research that this puts up, psilocybin is the focus of the current resurgence of research. As a result, it's a front-runner to become a licensed medicine, just behind MDMA (which we will look at in the next chapter).

## PSYCHEDELIC EFFECTS OF THE CLASSICS

There are a whole range of psychedelic effects mediated at the 2A receptor. Taking a classic psychedelic will change your perception, cognition and emotion as well as your sense of time, place and self. Below are the main effects that people report.

1. Visual changes. These include distortion, intensification of colors, textures, contours, light, changes in perception of size and shape, objects moving. People report

seeing and feeling the world in "higher resolution." Elementary hallucinations, seeing moving patterns (like colored Christmas tree lights) as well as complex hallucinations, which include seeing people, cities, galaxies, animals, plants or even gods.

2. Auditory changes in appreciation, sound and timbre. Synesthesia—changes of visuals led by changes in sound or music.

3. Changeable moods and emotions, from awe, wonder, euphoria, bliss, calm, joy, fun, excitement, love through to anxiety, fear and terror (especially in people with trauma-related mental health problems).

4. Changed perception in your body—of your size and shape—as well as where you are in time and place. Out-of-body experiences.

5. Altered or absent sense of self. This goes from seeing yourself in a new way right up to ego dissolution, the feeling of merging into your surroundings.

6. Reliving and reevaluating important memories. Deep introspection and psychological insights.

7. Enhanced positive emotions toward other people and your surroundings. With others, those might be trust, empathy, bonding, tenderness, forgiveness.

8. Cognitive changes such as increased creativity with shifts in thinking or putting ideas together in a new way, problem-solving.

9. Spiritual or mystical experiences. These can be related to a specific religion, such as seeing God or religious symbolism, or, more generally, seeing the connectedness of the universe.

10. Perceived contact and interaction with other creatures or entities.

Some of the classics stimulate receptors in the gut too. That's why it's common to experience nausea and vomiting after taking them. For example, at a traditional ayahuasca ceremony you're given your own bucket because vomiting or "purging" is often part of the experience.

Classic psychedelics have other negative effects too, including muscle spasms, temperature fluctuation, seizures, dizziness, paranoia and panic attacks, anxiety, confusion and inability to concentrate. Although some of these are likely due to changes in neurons and brain circuits that result from stimulating 2A receptors, some are probably caused by activity at other serotonin receptor subtypes and possibly at other receptors such as dopamine and noradrenaline too.[1,2]

## TIME AND SPACE

If you've read descriptions of people's experiences taking psychedelics, you'll know that no trip is ever the same, even if the same person takes the same drug. When we have given the same person two trips of the same dose of the same drug, there has been no obvious relationship between the content of each trip.

Although we can explain the reason psychedelics produce visual and other hallucinations (see Chapter 4), we still have almost no idea what determines the content of the psychedelic experience. Michael Pollan has a suggestion, in his book *How to Change Your Mind*. Specifically, that the content comes from the person's thoughts and intents immediately before the trip. For example, when Pollan was using a computer to explore the visual impact of a trip he found himself stuck in the hard drive of his computer. When he was in the garden of his dead parents' house, two old trees became his mother and father.

In our depression studies (see Chapter 5), the content people often engaged with was their past traumatic relationships, but we guided them to do this.

People assume each of the classics gives different psychedelic effects, but most of the variation doesn't come from differences between these drugs. If you took an equivalent amount of each of the classics so that each had the same interaction with the same number of receptors in the brain, they would all have the same effects. The variation comes partly from the dosage, and partly from how the drug is taken. One study tested doses of LSD from 25mcg to 200mcg. The bigger the dose, the more effects the subjects reported.[3]

The method of taking the drug governs how fast it gets into your brain. Faster entry means more rapid changes in consciousness. Your brain is always adapting to its inner and outer environment, trying to maintain balance. The faster the drug goes in, the less time the brain has to adapt, and the more different the experience is from normal consciousness.

For example, when a classic psychedelic such as DMT is injected or vaporized, it goes into the brain very fast, and so the trip is very intense—it can even feel like a near-death experience.[4] By the time the brain has realized something is happening to it, it's already been profoundly changed. When a psychedelic is taken orally, the brain receptors aren't challenged as fast. The brain has time to compensate, so the effects aren't as extreme. (There is an unexplained exception: when LSD is given intravenously, the effects tend to build up at the same speed as taking it orally.)

So while DMT and Toad (see below) have a reputation for being the most intense of the classics, catapulting people into another dimension, this is likely to be mainly due to them usually being smoked or snorted.

The classic psychedelics do vary in how quickly they leave the brain, i.e., how long the effects last. For example, a smoked DMT trip is 10–20 minutes, but an LSD trip is around 12 hours. We know some reasons why this might happen but not all. For example, we know that DMT and Toad are rapidly broken down in the blood, and that an LSD molecule has an extra part that binds it strongly to the receptor and so gives it a much longer-lasting effect.[5]

The length of the trip also depends, to some extent, on dosage. In the LSD dosage study above, as the dose increased the average duration also increased, from 6.7 hours to 11 hours.[6]

Another variable that can change the effects is your genetics. Some people don't have any kind of experience after taking a psychedelic. In one study at Imperial College, we injected 20 people with 75mcg of LSD.[7] Two out of the twenty reported that the drug didn't have any significant effect. At the time, we were looking at their brains under a scanner, so we could see that they weren't just denying the effect; they also had a lack of changes in the brain. It could be that some people have a genetic variation in their receptors that stops the drug from working. Or they may be metabolizing the drug faster, and so lessening the effect, or less of it got into the bloodstream for some other reason (we couldn't measure levels in the blood).

Finally, the course of a trip also depends on the set (the mindset) and setting (the surroundings), which are discussed in detail in Chapter 13.

# A GUIDE TO THE CLASSICS

## LSD (D-LYSERGIC ACID DIETHYLAMIDE OR LSD-25)

Out of all the classics, this one has probably the worst reputation. In the 1960s anti-LSD propaganda told—mostly fantastical—stories of fried brains, traumatic flashbacks and psychotic breakdowns, addiction, cancer, birth defects and people jumping out windows because they thought they could fly. Aimed at the parents of middle-class, white teenagers and young adults, it told them that LSD was turning their children into addicts and dropouts.[8] And it promoted the fear that LSD was ruining a generation and society.

LSD was demonized for encouraging young people to "turn on, tune in, drop out." This was the counterculture call to arms, coined by the self-styled prophet and spiritual teacher Dr. Timothy Leary when he was speaking to the crowd at the Human Be-In, a gathering of 30,000 hippies in Golden Gate Park, San Francisco. Around that time, he also said: "The next six years will be a period of dramatic change in regard to the social acceptance of such psychedelic drugs as marijuana and LSD. By 1970 we'll have our first LSD congressman and our first marijuana-smoking judges."[9] LSD was banned in 1967.

# WHAT IS A PSYCHEDELIC?

In the early 1960s, Leary was a clinical psychologist at Harvard. He was one of the many scientists and psychiatrists of the time who hoped that LSD would be the key both to a greater understanding of human consciousness (for more, see Chapter 9) and a revolutionary treatment for mental health conditions, in particular addiction (see Chapters 7 and 12).

**How is LSD taken?**
Orally. In the 1960s, people usually took the colorless, odorless solution dropped on a sugar cube. Now, it often comes soaked into square paper blotters, which are then dried and chopped up into hundreds of doses, or as tiny pills (microdots) or gelatin pieces.

**How long does LSD last?**
Comes on in half an hour, lasts 12 hours, although the effects can wax and wane over this period.

The Swiss chemist Albert Hofmann created LSD in 1938 while working at the pharmaceutical company Sandoz. He later named it his "problem child." Hofmann synthesized LSD from a toxic fungus, ergot. At the time, ergot compounds had been used for many years to treat migraine and promote the contraction of the uterus to stop bleeding after the placenta is delivered.

The myth that's grown up around LSD is that Hofmann discovered it by chance. But the truth is, he was following the standard process of medicinal chemistry, of taking a molecule and refining it in different ways to see if it might be a useful medicine that could be patented. This is the same process that made, for example, aspirin from willow bark.

Hofmann himself wrote: "LSD was not the fruit of a chance discovery, but the outcome of a more complex process that had its beginnings in a definite concept, and was followed up by appropriate experiments, during the course of which a chance observation served to trigger a planned investigation, which then led to the actual discovery."[10]

There was, however, one element of chance. When Hofmann remade a batch of LSD in 1943, he ingested some by mistake and so became the first person ever to experience an acid trip.[11]

He noticed some odd effects. "I was seized by a peculiar restlessness associated with a sensation of mild dizziness. On arriving home I lay down and sunk into a kind of drunkenness which was not unpleasant and which was characterized by extreme activity and imagination. As I lay in a dazed condition with my eyes closed (I experienced daylight as disagreeably bright) there surged upon me an uninterrupted stream of fantastic images of extraordinary plasticity and vividness and accompanied by the intense, kaleidoscopic play of colors. The condition gradually passed off after about two hours."[12]

He decided to test it properly on himself, also standard practice at the time. And so, on April 19, 1943, he took 250mcg, assuming this would be a small dose. But as we now know, and as Hofmann quickly found out, an unusual characteristic of LSD is that compared to other drugs you need very, very little to do a lot. It turned out that 250mcg is a very high dose.

Forty minutes later, he wrote: "Beginning dizziness, feeling of anxiety, visual distortions, symptoms of paralysis, desire to laugh." And so began a seriously strong, bad trip.[13]

Because it was wartime and there were no cars, Hofmann had to cycle home. He was accompanied by his assistant—

fortunately, because the journey sounds hairy. "Kaleidoscopic, fantastic images surged in on me, alternating, variegated, opening and then closing themselves in circles and spirals, exploding in colored fountains, rearranging and hybridizing themselves in constant flux."[14] Later, he had to be seen by a doctor. But slowly, as the hours passed, his bad trip turned into a good one, and by the next morning he was describing feeling more alive than ever, seeing the world and nature anew.

Now, April 19 is Bicycle Day, when people all over the world celebrate Hofmann's discovery and the intervening 80 years of LSD-charged transformation of minds, the arts and history.

Hofmann's assessment was that as LSD alters consciousness, it would make a great tool to discover more about the mind. Sandoz Laboratories patented LSD, as Delysid, and sent it out free to a whole range of psychiatrists and researchers around the world to discover more about how it could be used. Before LSD was banned in the late 1960s, it was the most-used drug in more than 1,000 scientific papers on psychedelics, based on their administration to around 40,000 people.[15]

Since the revival of research in the 1990s, there have been studies on end-of-life anxiety, pain and depression, and addiction (for more, see Chapter 7). However, the amount of research on LSD has, so far, been more limited than that on psilocybin. Possibly researchers have been put off doing trials on LSD by the historical animosity to it or lingering fears around it. They might assume these would make the permissions or the funding needed less forthcoming. Or it may be because psilocybin (see below) produces a shorter trip that better fits the timescale of studies. The other take on it is a joke I make to scientists: psilocybin doesn't come in for as much public criticism as LSD because no politician or journalist knows how to spell it.

## PSILOCYBIN

Mushrooms that make psilocybin have been called "magic mushrooms" since the moment they hit Western consciousness. The name was coined in a 1957 article in *Life* magazine, "Seeking the Magic Mushroom." The author was R. Gordon Wasson, a banker at J. P. Morgan in New York who, like his wife, Valentina, was a keen amateur mushroom hunter. Wasson heard about the secret mushroom ritual or *velada* (vigil) of the Mazatec people of Oaxaca state in Mexico, then spent two years searching for one to take part in. He eventually found a *curandera* (healer) called Maria Sabina, who allowed Wasson to attend one of her overnight ceremonies.

Valentina Wasson went to see Sabina too: her own story came out a week later in a Sunday supplement called *This Week*. She wrote: "My mind was floating blissfully. It was as if my very soul had been scooped out and moved to a point in heavenly space, leaving my empty physical husk behind in the mud hut. Yet I was perfectly conscious. I knew now what the shamans meant when they said, 'the mushroom takes you there to the place where God is.'"[16]

### How is psilocybin taken?

Orally. You can eat the mushrooms fresh or dried, or grind them and add hot water to make a mushroom tea. They can also be infused into a tincture. The chemical in the mushroom

—psilocybin—is very stable. When you consume it, it's metabolized in the stomach and blood into the active ingredient, psilocin.

In the Netherlands, the truffles—the rhizomes or underground part of the mushrooms—are legal and are taken in the same way as mushrooms. For research, we use a synthetic version of psilocybin because the dose can more easily be standardized.

### How long does psilocybin last?

It comes on in half an hour when you take it orally and lasts five to six hours. In our first experiments, we didn't have the funding to buy enough psilocybin to give it orally, so we gave it intravenously, which makes it come on within minutes.

Wasson did not conduct himself with distinction in Mexico. He explained: "The mushrooms are not used as therapeutic agents: they themselves do not effect cures. The Indians 'consult' the mushrooms when distraught with grave problems." And so when Sabina was reluctant for Wasson to attend the velada, he persuaded her by inventing a problem with his son.[17]

Sabina eventually agreed that Wasson could participate in the velada on the condition he kept it secret. Wasson did use a pseudonym for her in his piece in *Life* magazine, but he later divulged her real details. The Wassons' articles were read by millions; the secret was definitely out. The curandera Sabina's remote mountain town, Huautla de Jiménez, became overrun by spiritual tourists and hippies who wanted to experience the velada. There's a long list of celebrity attendees who reportedly made the journey, including John Lennon, Bob Dylan, Keith Richards and Walt Disney.

Sabina's story doesn't have a happy ending: over a few years, she was raided by Mexican authorities and her son was murdered, then her home was burned down. She eventually died of causes related to malnutrition.[18]

Wasson took some of the Mexican mushrooms and gave them to Hofmann, who deduced that the active ingredient was psilocybin. Sandoz, Hofmann's employer, then began manufacturing and marketing psilocybin as a therapeutic medicine alongside LSD; they called it Indocybin.

We now know there are at least 200 species of magic mushrooms. There is evidence that they have been used in human societies for thousands of years, the earliest known being Algerian cave pictures dating from 5000 BC of shamans holding mushrooms.[19]

They're found all over the world, from northern Canada to the southern tip of New Zealand. The one that's most common in the UK is liberty cap. In the U.S., it's *Psilocybe cubensis*, also called golden cap. We don't know why mushrooms make psilocybin; it may be either to defend against being eaten or to encourage it. But psilocybin is likely to have some function, simply because it occurs in so many species.

Out of the classic psychedelics, psilocybin is the most likely to be the first to become a licensed medicine. In the U.S., this could be as early as 2024. Following promising preliminary trials, the FDA (U.S. Food and Drug Administration) has given it Breakthrough Therapy status, which can speed up regulatory approval, for both treatment-resistant depression and major depressive disorder.

Psilocybin is being researched and trialed for multiple conditions. The list includes more types of depression, such as in cancer patients and people with Alzheimer's disease, drug dependence (alcohol, nicotine, cocaine, opioids), anorexia nervosa and obsessive-compulsive disorder. And also for pain: chronic pain syndromes and headaches (including cluster and migraine). (See Chapters 5, 7 and 10 for more.)

Psilocybin is currently banned in most countries. There are exceptions, including the Netherlands (as truffles), Jamaica and the states of Colorado and Oregon, U.S. And places where it's been decriminalized include Spain and Portugal as well as Oakland and Santa Cruz in California plus Ann Arbor, Michigan, and Washington DC. There are psilocybin retreats in Jamaica, Costa Rica, Spain, Portugal and the Netherlands. From July 2023 psilocybin will be an approved medicine for treatment-resistant depression in Australia.

## DMT (N,N-DIMETHYLTRYPTAMINE)

DMT is best known as an active ingredient in ayahuasca, which is also called yagé. It's a thick, red-brown tea from the Amazon basin that's used in shamanic ceremonies. It's also the official sacrament of two churches in Brazil, Santo Daime and União do Vegetal, that have spread around the world; it replaces the blood of Christ in their quasi-Catholic masses. In 2006, the U.S. Supreme Court ruled that members of the União do Vegetal church could use ayahuasca as their sacrament in that country also.

DMT comes in other forms too: a South American smokable plant-leaf version of ayahuasca, often called changa, and a synthesized version of DMT that comes as a white or yellow-pink crystallized powder.

### How is DMT taken?

DMT is broken down very fast in the gut and the liver by an enzyme called monoamine oxidase, which is why changa is smoked, and synthesized DMT is smoked or vaped. It's also why ayahuasca is made from two plants: usually the *Psychotria viridis* shrub and the *Banisteriopis caapi* vine. The first contains the DMT, and the second contains substances that inhibit the enzyme (monoamine oxidase inhibitors or MAOIs). MAOIs stop DMT from being broken down so fast, giving it time to get into the brain.

### How long does DMT last?

DMT is very fast-acting, with effects within seconds when it's smoked or injected intravenously. People often describe it as taking them into a different dimension, often with strong visual hallucinations and seeing entities or creatures. Smoked or intravenous, the duration of its effect is about fifteen minutes.[20]

Ayahuasca comes on more slowly and can last up to five hours, although at ceremonies it's often given every few hours.

Ayahuasca isn't a party or festival drug. For most people who take it, it's not about getting off their head but about getting to a place where they can work to become a better person. In the Santo Daime church, ceremonies are called "works" and they include group dancing and singing of hymns, and taking ayahuasca every two or three hours. Services can last up to 12 hours.[21]

In 2016, I was the scientific adviser for a BBC documentary, *Getting High for God?* In one episode, the presenter and comedian Mawaan Rizwan traveled to Brazil to join an ayahuasca church.[22]

WHAT IS A PSYCHEDELIC?

The producers asked me: "When the presenter is given a cup of ayahuasca, how much should he drink?" I said, "Drink a cup, if it's what others drink." The producers came back and said: the lawyers want you to be more specific about Mawaan's safety. So I suggested he drink half a cup. The lawyers seemed happy with that. You'll be pleased to know he survived to tell his tale (see below)!

Another name for ayahuasca is "la purga" (the purge) because people often feel nauseous, vomit and have diarrhea (these are considered to be cleansing).

Mawaan said his trip started with 20 minutes of feeling nauseous, then the ayahuasca hit. "As the ceremony leaders played the tabla and sang ritualistic songs, I closed my eyes and became totally engrossed in a spiral of thoughts and visions. I laughed, I wept and I saw things from my childhood that made me extremely uncomfortable."[23] He said some of the visions were distressing and confronting, but the shamans told him to go deeper into them, not run away from them. The trip went on for five hours, then he slept. When he woke up the next morning, he felt he'd found answers to many of the questions he'd had about life.

It's thought that ayahuasca—which translates as "vine of the soul" or "vine of the dead"—has been used by indigenous people of South America for thousands of years. In the sixteenth century, the conquistadors saw it as the influence of the devil, because people who took it saw serpents. This is a classic example of a cultural misunderstanding; in Christian culture, the serpent is a representation of evil but in these indigenous cultures, it was and still is a sacred symbol.[24]

The past ten years have seen a boom in spiritual tourism, people traveling to the South American jungle to experience ayahuasca ceremonies. This has brought problems to the area; there have been reports of sexual harassment and some deaths of Westerners too. This may be down to a combination of troubled people seeking help, and a lack of safety and safeguarding at retreats. Also other plants are sometimes added to enhance visual effects, and these sometimes contain harmful ingredients. The tea is also increasingly being offered at retreats in Western countries and in less ceremonial settings. The Canadian psychiatrist Dr. Gabor Maté has described using ayahuasca in treating patients for trauma and addiction.[25] There's one published clinical study from Brazil of ayahuasca in depression that showed good effects.[26]

My team is currently helping the pharmaceutical company Small Pharma to research DMT for major depressive disorder (MDD). Preliminary results were very positive; a single 20-minute infusion with psychological support before and after produced significant reductions in depression scores.[27]

There's also commercial research going on in Switzerland by MindMed that is looking for more indications.[28]

## TOAD OR 5-MEO (FIVE-METHOXY-N, N-DIMETHYLTRYPTAMINE, OR 5-MEO-DMT)

It sounds improbable that a psychedelic could come from milking a toad. But 5-MeO comes from the white, milky, poisonous secretions of the glands of the Colorado River or Sonoran Desert toad (formerly called *Bufo alvarius*, now *Incilius alvarius*). 5-MeO is also found in some plants and is the active ingredient of *yopo* or *virolo*, a shamanic plant snuff from

South America. Yopo has a documented history of around 3,000 years and is still used today. 5-MeO was first synthesized in 1936 and first isolated from a plant in 1959.

### How is 5-MeO taken?

When the toad secretion dries, it forms off-white crystals that are then smoked. Yopo is made by toasting and then grinding the seeds of the *Anadenanthera* plant. They're then blown up the nose by a shaman. The synthetic version is a powder.

### How long does 5-MeO last?

It's a very intense trip that comes on within a minute and lasts 20 minutes on average when smoked, and slightly longer when snorted.[29]

The Sonoran toads are the largest native toad in the U.S., up to 7.5 inches (19cm) long, with green-gray, fairly smooth skin with a few warts, and a pale belly. They live in the southwestern U.S. and northwestern Mexico. They're carnivorous, eating small animals and insects. When disturbed, they secrete a toxin containing 5-MeO from glands on the back of their head.

Most likely, the toxin evolved to scare away predators, who might want to avoid a second bout of nausea and/or dizziness. Or it could have evolved to actually kill predators: the toxin is strong enough to kill a full-grown dog. It's four to six times more potent than DMT.[30]

The experience, which comes on very quickly, is said to be so intensely psychedelic that it's been dubbed the "spirit molecule" or "God molecule." The boxer Mike Tyson described it like this, on a podcast presented by Joe Rogan: "I look at life differently, I look at people differently. It's almost like dying and being reborn...you're submissive, you're humble, you're vulnerable—but you're invincible still in all."[31]

The first report of toad secretions being smoked was not until 1983, by an artist called Ken Nelson. He dried the secretions on his windscreen, smoked it and said he was changed forever. But some practitioners claim that Toad has a much older, indigenous history, citing ancient art that includes images of frogs.

Despite it being made illegal in the U.S. in 2011, the number of users is growing fast. One key player in this is the controversial Mexican physician Dr. Octavio Rettig, whose 2014 TEDx talk describes his work getting people off crystal meth addiction using Toad. This use of 5-MeO in addiction is now under research.

The use of 5-MeO is mainly unregulated and there have been multiple reports of people taking Toad in an unsafe way and dying. Some practitioners have also been accused of sexual assault, fraud, serious mistreatment and psychological manipulation.[32] There are also warnings that poaching and trafficking are putting the toad itself at risk of population collapse. As a result, some practitioners are now using synthesized 5-MeO.[33]

So far, only a handful of studies have been done on 5-MeO. Johns Hopkins University School of Medicine surveys of people who've used Toad suggest it may be useful in anxiety and depression,[34] as well as improving satisfaction with life.[35]

At Imperial College, we are currently conducting some of the first dose-finding studies with 5-MeO in preparation for brain imaging work to see how it compares with the other classic psychedelics.

## MESCALINE (3,4,5-TRIMETHOXYPHENETHYLAMINE)

The two main sources of mescaline are the wachuma or San Pedro cactus (*Trichocereus pachanoi*/*Echinopsis pachanoi*), originally from South America, and the peyote cactus (*Lophophora williamsii*), from Mexico and North America. Effigies made of peyote cactus have been found in Texas that date back to 4000 BCE, showing that it's been taken for at least 6,000 years.[36] In South America, archaeologists have found pestles and mortars for grinding the cactus flowers, belt pouches for carrying the powder, and ceramic straws for snorting it. It's thought to have been used in religious ceremonies and to treat snakebites, burns, wounds, rheumatism, toothache and fever.[37]

From the sixteenth century, the Spanish conquistadors tried, as they also did with ayahuasca, to eliminate the use of mescaline, damning it as *raíz diabólica* or "root of the devil." In 1620, the Roman Catholic Church passed ecclesiastical laws against peyote, calling it "an evil to be rooted out in the New World." People could be tortured and even killed for using it.[38]

### How is mescaline taken?

Mescaline is taken as a white or brown powder, often in a capsule, or as a liquid. There's also a synthetic version.

### How long does mescaline last?

The effects come on within one to two hours. It lasts up to 12 hours.

In North America, mescaline use was mainly confined to where it grew in Texas, until the nineteenth-century persecution of the Native American peoples. People were forced into reservations and moved from reservation to reservation. The loss of traditional lands, together with the destructive introduction of alcohol, destroyed social cohesion. As a response, even tribes who didn't have a history of peyote use adopted it, and the peyote ritual became a way to strengthen community solidarity in the face of terrible deprivations.

For the next hundred or so years U.S. prohibitionists attempted to ban mescaline, with varying degrees of success. In 1880, for example, the U.S. Congress passed laws saying that anyone caught in possession of peyote would be locked up for 90 days and their government rations stopped. The laws also outlawed indigenous religious ceremonies. The Native American Church (NAC) was established in 1914 as an act of resistance and named peyote as its sacrament. The NAC grew and finally, in 1994, Native Americans gained a legal right to use and possess peyote for "traditional ceremonial purposes," according to the American Indian Religious Freedom Act Amendments, passed by President Clinton. With the exception of the Native American Church, peyote and mescaline remain illegal in the U.S. The church has grown and now has an estimated quarter- to a half-million members.[39]

The peyote cactus is under threat. It's very slow growing, taking ten years to grow from a seed to a size at which it can be used.[40] There are now protected plantations so Native Americans can have guaranteed access. The San Pedro cactus is a more sustainable source.

Mescaline was the first psychedelic to be identified chemically, in 1897, by German chemist Arthur Heffter. At the turn of the century it became the fashionable tool for exploring consciousness, used by scientists, artists and thinkers. In the 1930s, Jean-Paul Sartre had a bad trip, as described by Simone de Beauvoir: "The objects he looked at changed their appearance in the most horrifying manner: umbrellas had become vultures, shoes turned into skeletons, and faces acquired monstrous characteristics..."[41]

One of the best-known accounts of trying mescaline is undoubtedly Aldous Huxley's; he describes it in *The Doors of Perception*, writing that it shows "the miracle, moment by moment, of naked existence" and his surroundings as "just of light, of everything flooded with light...a kind of luminous living geometry."[42]

That book was one of the sparks that lit the fire of the psychedelic 1960s. Huxley became a great advocate for psychedelics, introducing them to Timothy Leary and Beat writer Allen Ginsberg among others. He became a cult figure, doing lecture tours around the U.S., putting both the experience and the science of psychedelia into the public consciousness.

It was Huxley who created our modern way of thinking about psychedelics, as a scientific tool for exploring consciousness but also for learning about spirituality. He also helped coin the word "psychedelic," in his correspondence with the psychiatrist Humphry Osmond.

At one point in Michael Pollan's mescaline trip, described in his book *This Is Your Mind on Plants*, he describes his new appreciation of "the immensity of existing things" and that the peak of the trip (and here he quotes Huxley) is "wonderful to the point, almost, of being terrifying." And mescaline doesn't produce a short trip. "Mescaline goes on and on—it is, assuming you're enjoying it, the most generous of psychedelics—and I settled in for the twelve-hour ride."[43]

Mescaline does differ from other classic psychedelics, as it's a molecule called a phenethylamine rather than a tryptamine, making it part of the same chemical family as amphetamines and MDMA. However, its psychedelic effects come from binding to the 2A receptors as the other classic psychedelics do.

A recent study found that peyote users in the NAC were more psychologically healthy than lifetime abstainers when scored on anxiety, depression and psychological distress.[44]

And in a survey of people who'd used mescaline, people reported finding it effective for depression, anxiety, PTSD and drug dependence.[45]

## THE NBOME PROBLEM

NBOMes are novel psychedelics, created in the 2000s as possible new medicines. But they were discovered by psychonauts and slipped into recreational use. This led to them becoming illegal in many countries. They're sold as N-Bomb, Solaris, Smiles and Wizard, among other names. They combine stimulation of serotonin receptors with effects on serotonin uptake and release. This makes them exceptionally potent and gives them broader effects than the classics. After some deaths were reported, most were banned, which is a good thing as many of them are considerably more toxic and harmful than the classics. There is very little animal data and virtually no formal research in human beings.[46]

In one UK case, a 26-year-old IT specialist, Richard Phillips, suffered catastrophic brain damage after taking N-Bomb. His temperature soared and he suffered seizures and multiple organ failure.[47]

At the time, some people called for the censorship of the formulas of NBOMes. But although there have been terrible consequences from the formulas being in the public domain, censorship of science is problematic. If scientists can't publish, how can the field move on? What if what you can't publish could be clinically useful? In fact we use an NBOMe called Cimbi-36 as a PET tracer for the 2A receptor because it's so very potent that we can see binding even with minuscule (well below psychedelic) doses.[48]

NBOMes are a problem because they're so potent and very toxic. But the real problem is that they were legal while safe and known alternatives, such as LSD, were not. This is another case of prohibition leading to a legal alternative that then turns out to be more dangerous. For example, the prohibition of cannabis led to a market for the (then) legal drug spice. The best way to protect the public, in this case, is to inform them of the danger of NBOMes. But an even better way would be to set up a regulated market in safe and known drugs, so there's no market for toxic alternatives.

# Chapter 3

# AN INTRODUCTION TO THE NON-CLASSICS: MDMA, KETAMINE, IBOGAINE

**WHAT UNITES THIS** disparate group of compounds is that they are psychedelics—or at least psychedelic-like—because they produce changes in mental states, specifically in perceptions and thought processes. They aren't classic psychedelics because they don't work on the serotonin 2A receptor. Some, such as ketamine and salvia, produce profound alterations in consciousness that have some similarities to those produced by the classics. Others, such as MDMA, share therapeutic effects that act through the serotonin system. I could have included more compounds, but I chose these because they're the ones currently of most interest in research and therapy.

## KETAMINE

Developed in the 1960s as a sedating and pain-killing anesthetic, ketamine is still widely used by both doctors and vets. This explains why it's sometimes called "horse tranquilizer": it is one. The special value of ketamine is that, unlike other anesthetics, it doesn't suppress breathing. This makes it very safe and very useful in situations where there isn't an anesthetist with machines to ventilate the patient. And so it can not only be used on the battlefield to help injured soldiers but in the farmer's field to treat horses. When one of my team, who's an anesthetist, worked at a South Sudan refugee camp, she said the resident medical lead, a former paramedic, was more than qualified to administer ketamine for operations.

As you'll see in Chapter 5, it's already being prescribed off-label as an antidepressant.

### How is ketamine taken?

Street ketamine is a white or transparent powder of crystals that's usually snorted. In clinic and studies, we give it as an intramuscular injection (i.m.) or intravenously (IV). In 2019, Johnson & Johnson brought out a nasal spray version called esketamine, brand name Spravato, which has been licensed for treating depression in a number of countries including the U.S. and UK.

### What does ketamine feel like?

Depending on how much is taken, an overwhelming feeling of relaxation, body numbness and separation from your environment, confusion, nausea, euphoric with distorted sights and sounds.

### How long does ketamine last?

Taken nasally as a powder, it takes 5 to 15 minutes to come on and lasts around an hour.

Ketamine was developed from the anesthetic phencyclidine (PCP) because of PCP's serious side effects, which include lengthy delirium.[1] Like alcohol, PCP (its most common street name is angel dust) stops the cortex—the brain's controller—behaving rationally. It also makes people extraordinarily strong, which explains the stories of people high on PCP fighting multiple police officers or lifting cars.

Ketamine turned out to be an excellent alternative to PCP, being both a good painkiller and simple to administer as previously explained. It gained the nickname of the "buddy drug" because it was so safe, even soldiers without medical training could use it. In the Vietnam War, combatants were given a pack of ketamine and morphine injections. If their fellow soldiers were injured, they gave them this combination to help them dissociate and reduce their agony.

Ketamine is increasingly being used as a painkiller; it's given intravenously. More recently, it's been given subcutaneously via a pump for chronic pain syndromes.

Ketamine became a street drug around the 1990s but really took off in the 2000s. At first people used it post-rave to help them come down from MDMA, then it became a rave drug in its own right.

Another reason it became popular is that it's a relatively cheap hit. One of the places ketamine first took off for non-medical use was Bristol. There's an urban legend that Banksy, the Bristolian underground artist, once said: "The best Friday night in Bristol is eight pints of cider and two lines of ketamine, then I throw up over Bristol bridge into the river."

I was the chair of the ACMD technical committee in 2004 when we recommended that ketamine be made illegal and put into Class C. At the time, it was being manufactured in bulk in India and being imported in huge drums, labeled as "rosewater" and "massage oils." There were an increasing number of people becoming dependent, and some ten or so a year were dying. More were ending up with long-term health problems, such as bladder damage and brain damage (see Chapter 13).

In 2015, China put pressure on the UN Commission on Narcotic Drugs (UNCND) to make ketamine a Schedule 1 drug globally, in order to stop people in Hong Kong from abusing it. This would have classified it as having no medical use and so would have stopped it being used as a medicine.[2]

The charity I set up, Drug Science, lobbied the UN not to go ahead with this. For a start, the idea that banning ketamine would stop recreational use is ludicrous, given that similar bans on heroin and cocaine have not.

Banning ketamine would also deny a useful medicine to so many people. This happened with opioids; many countries such as India and China blindly followed UN guidance to ban all strong opioids. The result is that 80 percent of the world's population do not have access to opioid painkillers, one of the great socio-medical scandals of the past century.

If the UN had passed the recommendation, and the UK had implemented it, every doctor and hospital wanting to use ketamine would need a license to do so. Quite apart from the paperwork for obtaining one, which takes a year, the cost would be around $7,000. This would effectively stop the clinical and research use of ketamine. Fortunately, lobbying from doctors, vets and scientists meant the rescheduling did not go ahead. Subsequent research into the use of ketamine for depression and dependence has shown just how important this decision was (see Chapter 13).

Russia is one of the countries where ketamine is illegal. Following lobbying by the actor Brigitte Bardot, in 2004 Putin made a special concession to allow it to be used by vets.

Ketamine primarily works by dampening down the glutamate system, the excitatory, wake-up system of the brain. The resulting psychedelic effects are very different from those of the classic psychedelics. People describe feeling dissociated, detached, in a dreamlike state, confused, anxious or nauseous. It changes perception of time and space and produces hallucinations.

The higher the dose, the closer you get to the anesthetic effect of no longer being able to move or stand. This is the so-called K Hole, where someone will appear slumped and unconscious while they're likely having an out-of-body, hallucinatory experience. The K Hole experience is on the way to the experience of being anesthetized, but you're not so out of it that you can't remember it afterward.

Even though ketamine affects different brain mechanisms than the classics, when you look at brain images of people dosed with it you can't tell it apart from psilocybin or LSD.[3] What you see on the scanner is disorganized, fragmented, unsynchronized firing of the brain. Because ketamine has similar effects on the brain but is already proven to be a very safe medicine, it's already being used to treat depression in clinics in the UK and the U.S. and many European countries.

In New Zealand, psychopharmacologist Professor Paul Glue is trialing a long-acting ketamine pill for maintenance therapy for people with recurrent depression and anxiety. And the spray esketamine is now widely prescribed for treatment of depression in the U.S. (more in Chapter 5).

Disclosure: I am chief research officer at Awakn Life Sciences, helping develop psychedelic-assisted psychotherapy. Currently we are working with US ketamine clinics to provide the skills they need to treat these disorders.

## MDMA (3,4-METHYLENEDIOXYMETHAMPHETAMINE)

Although MDMA didn't become a party drug until the late 1980s, it was first made 100 years before, in the 1880s. In 1914, it was patented by the German drug company Merck. The myth around its invention is that it was developed to control the appetites of German soldiers. The truth is nothing as dramatic: MDMA wasn't even patented for itself but as part of the manufacture of a new drug to control bleeding.[4]

### How is MDMA taken?
As a pill (ecstasy) or a powder. In the lab as a capsule.

### What does MDMA feel like?
Depending on how much is taken, a general feeling of well-being and friendliness, a lifting of anxiety and fear, a euphoric rush, distortion of perceptions.

### How long does MDMA last?
Six to eight hours.

MDMA is part of the amphetamine family, a group of drugs that was first used to save the lives of people with asthma. Before their invention, one of the only drugs available that could dilate the bronchi during an asthma attack was ephedrine, made from the plant ephedra. After the pharmaceutical company Boehringer synthesized methyl amphetamine, or crystal meth, it was used as a treatment for asthma for the next 100 years. Even as late as 1972, when I was a medical student, I injected meth intravenously into a man having a serious asthma attack to help him stay alive. (Now there are more targeted drugs for asthma, but meth is still a prescribable medicine for narcolepsy.)

It took until the 1970s for MDMA's particular effects to come to light. It was rediscovered by medicinal chemist and psychopharmacologist Alexander "Sasha" Shulgin. He had a Schedule 1 license that allowed him to possess, analyze and make psychedelic drugs, in order that he could give expert testimony at drug trials. He wrote the book for drug law enforcement—literally—it's called *Controlled Substances: A Chemical and Legal Guide to the Federal Drug Laws*.[5]

However, Shulgin's true motivation wasn't to help law enforcement but instead to explore the limits of human consciousness and psychoactive chemistry. To that end, over the course of his career he invented more than 100 novel psychedelics, as well as testing out thousands of existing formulas.

Shulgin had his own research laboratory at his house, The Farm, in the Bay Area, San Francisco. In pictures, his workspace is the epitome of a mad scientist's lab, filled with a floor-to-ceiling framework of suspended pipes and glassware.[6]

His modus operandi was to test new substances on himself first, at a very low dosage. If that went well, he'd increase the dosage every other day. Next, he'd try them out on his wife Ann, a therapist, and a close group of friends. "A lot of the materials in Schedule 1 are my invention," he said. "I'm not sure if it's a point of pride or a point of shame." In a *New York Times* article, Ann estimated that she'd had more than 2,000 psychedelic experiences; Shulgin thought he'd had more than 4,000.[7]

Shulgin trialed MDMA in 1976, after hearing anecdotes about its peculiar qualities. He quickly realized it could help people be more emotionally open, empathetic and compassionate, both to others and themselves. In a paper written soon afterward, he and medicinal chemist and pharmacologist Professor David Nichols wrote that MDMA brings about "an easily controlled altered state of consciousness with emotional and sensual overtones." More casually, Shulgin described it as a "low-calorie martini."[8]

Along with Ann and Leo Zeff, another therapist, Shulgin began to extol the use of MDMA in therapy and couples counseling. Within the industry, it quickly gained a reputation for having the ability to accelerate progress and enable breakthroughs in therapy.

The use of MDMA quickly spread from West Coast therapy circles to the East Coast, and from there on to party users in Texas, Ibiza and beyond, launching the rave scene. From the late 1980s, it began to be banned in pretty much every country around the world.

Shulgin's response was to self-publish, in 1991, a book of recipes, descriptions and dosages for MDMA and 178 other substances, called *PiHKAL: A Chemical Love Story*. PiHKAL stands for Phenethylamines I Have Known And Loved, phenethylamines being the family of drugs that includes MDMA and other amphetamines.[9]

It wasn't that Shulgin supported indiscriminate use of these drugs, although publishing their recipes might make it appear that he did. "Go banging about with a psychedelic drug for a Saturday night turn-on, and you can get into a really bad place, psychologically," he wrote in *PiHKAL*.[10]

Rather, he thought this information should be out in the world. He commented, "Our generation is the first, ever, to have made the search for self-awareness a crime, if it is done with the use of plants or chemical compounds as the means of opening the psychic doors."[11]

The authorities didn't see things his way. In 1994, The Farm was raided by the DEA, Shulgin was fined $25,000 and his drug-handling license was taken away.[12]

But he persisted, publishing *TiHKAL: The Continuation* (which stands for Tryptamines I Have Known And Loved),[13] a compendium of the classic psychedelics and wider family.

Now Shulgin's original favored use of MDMA, as a tool for therapy, is being revived for treating PTSD. It looks as if it will be the first psychedelic to gain FDA approval as a medicine. (For more on the use of MDMA in therapy, see Chapter 8.)

## SCOPOLAMINE

Scopolamine was first extracted in 1880 from a plant in the nightshade family. It's one of the belladonna alkaloids, which like many poisons, can be useful in medicine. When I was a junior doctor, we'd give opium and scopolamine to dry up people's saliva so they wouldn't choke during operations. In fact, scopolamine is still used for motion sickness, and as patches for postoperative nausea. It's a banned substance in horse racing as it can improve a horse's breathing and heart rate.[14]

**How is scopolamine taken?**
In South America, as a powder called *burundanga* that's snorted. It's also sometimes added to ayahuasca mixtures to enhance their effects.

**What does scopolamine feel like?**
Confusion, dizziness, agitation, drowsiness, hearing voices or seeing things that don't exist.

**How long does scopolamine last?**
On its own, four to six hours.

Scopolamine became infamous in 1910 as the drug used by the American homeopath Dr. Hawley Crippen to poison his wife Corrine "Cora" Turner, a music hall singer. When Cora disappeared after a party, Crippen told friends she'd returned to the U.S. But Cora's friends became suspicious and tipped off the police, who began to investigate. Crippen told them that Cora had gone off with another man.

When Crippen disappeared a few days later, the police searched the couple's house. They found a body under the kitchen coal cellar, with no head or genitals, that was identified as Cora's. It contained a lethal dose of scopolamine.

A manhunt led to Crippen being discovered traveling as Mr. Robinson on a ship to Quebec, accompanied by his ex-secretary Ethel Le Neve, who was disguised as a young boy.

They couldn't be arrested until the ship docked. Daily updates from the ship as it crossed the Atlantic caused a press frenzy. At the trial, held at the Old Bailey, the jury agreed with the prosecution that Crippen had poisoned Cora with scopolamine. He was sentenced to death and hanged shortly afterward. Ethel was acquitted.[15]

(n.b. In 2007 DNA expert Dr. David Foran reexamined evidence from the body and concluded that the body couldn't have been Cora, one key reason being that it was male!)[16,17]

More recently, you may have heard of scopolamine as devil's breath. This is the South American powdered version of scopolamine—*burundanga* or *borrachero*—that's extracted from the *Brugmansia* plant (angel's trumpet) or from other nightshade plants. There have been (unproven) reports of gangs in Columbia and Ecuador, and in Paris, blowing burundanga into people's faces in order to stupefy them, then robbing them.

In 2021, a TikTok user's accidental encounter with scopolamine went viral. Singer-songwriter Raffaela Weyman, from Toronto, Canada, filmed herself and her best friend sniffing a yellow flower with a scent she described as "delicious." The flower was yellow angel's trumpet, which contains a variety of belladonna alkaloids including scopolamine.

"When we arrived at our friend's birthday, we both suddenly felt so f****d up and had to leave," Weyman said in her TikTok video.[18]

Her experience shows how scopolamine's psychedelic

effects are completely different from those of the classic psychedelics. "I went to bed and had my first sleep paralysis experience—I thought a human entered my room dressed in black and sat next to me injecting me with a needle that made me unable to talk or scream or move. I was just lying there making quiet moaning sounds," she told *Newsweek* magazine.[19]

Scopolamine also affects your memory. Oliver Sacks was a physician and neurologist who took almost every psychedelic, including Artane, a synthetic drug that's very similar to scopolamine.

In his 2012 book *Hallucinations*, he describes this experience.

"Go on," urged my friends. "Just take twenty of them—you'll still be in partial control." [three to four was a medical dose for Parkinson's].

So one Sunday morning, I counted out twenty pills, washed them down with a mouthful of water, and sat down to await the effect...

I was in the kitchen, putting on a kettle for tea, when I heard a knocking at my front door. It was my friends Jim and Kathy; they would often drop round on a Sunday morning. "Come in, door's open," I called out, and as they settled themselves in the living room, I asked, "How do you like your eggs?" Jim liked them sunny side up, he said. Kathy preferred them over easy. We chatted away while I sizzled their ham and eggs—there were low swinging doors between the kitchen and the living room, so we could hear each other easily. Then, five minutes later, I shouted, "Everything's ready," put their ham and eggs on a tray, walked into the living room—and found it completely empty. No Jim, no

Kathy, no sign that they had ever been there. I was so stag-gered I almost dropped the tray.[20]

Scopolamine offers a different door to exploring consciousness, but it's hardly been studied, possibly because it's so unpleasant. Sacks's brain on scopolamine created not only a fake experi-ence but also a fake memory of that experience, something he found shocking. Scopolamine has this effect because it blocks the acetylcholine/cholinergic system. This is the system that helps us to lay down memories and allows us to learn.

It's fascinating that it's possible to make memories about an event that never happened. It would be interesting to use imaging studies to find out what's going on in the brain in this state of dissociative altered consciousness. In research, scopol-amine has been used as an experimental model for memory problems in Alzheimer's.[21]

There have also been studies showing that scopolamine may have an antidepressant effect.[22] However, a large-scale trial of a new synthetic form of acetylcholine receptor blocker that looked promising has failed.[23]

## IBOGAINE

Ibogaine, extracted from the roots of *Tabernanthe iboga*, has been a sacred plant for centuries in west central Africa. In the Bwiti spiritual tradition, iboga root is still used for ceremo-nies, to find insights and to commune with ancestors and the forest. One of the Bwiti teachings translates as: "respect all of nature—plants, animals, humans, ourselves." The major Bwiti iboga ceremony is an initiation ritual that lasts for days, mark-ing the transition into adulthood.[24]

# AN INTRODUCTION TO THE NON-CLASSICS

**How is ibogaine taken?**
Orally, either by chewing the root bark shavings of the iboga plant or as a brown powder.

**What does ibogaine feel like?**
Out-of-body, dreamlike trance state with closed-eye hallucinations.

**How long does ibogaine last?**
Usually 24 hours but up to 72 hours, strongest in the first 8 to 12.

The first reports of ibogaine use came from Belgian and French missionaries. In the 1930s, it was marketed in France as a neurostimulant called Lambarène. This weak tincture of ibogaine was sold for "depression, asthenia [lack of energy], convalescence, infectious disease, [and] greater than normal physical or mental efforts by healthy individuals." It contained a very low dosage; you could say it was the Western world's first instance of commercialized microdosing.[25]

Ibogaine's long, powerful, dreamlike and disturbing trip can last for days—up to 72 hours—and includes a sense of ego death that can last for hours. In fact, it lasts so long that people often say they wonder if they're ever going to recover.

Very few people like taking it, but there's a lot of anecdotal evidence that it's very helpful for the treatment of dependence. It's been used widely since the 1980s for treating addiction and withdrawal, particularly from opiates but also cocaine.

Ibogaine's history as a rehab treatment is checkered. Its anti-addictive properties first surfaced in 1962, when a 19-year-old American heroin addict, Howard Lotsof, tried it and reported that his desire for heroin disappeared. He began a lifelong crusade to make ibogaine available as a medicine to treat dependence. However, in the late 1960s it was banned in the U.S., along with other psychedelics.[26]

Early studies initiated by Lotsof indicated that it helped around two-thirds of opiate addicts stop using. And in 1986, Lotsof managed to obtain U.S. patents for its use in heroin as well as cocaine, amphetamine, alcohol, nicotine and polysubstance addiction. However, the required safety and efficacy studies were never completed for it to be licensed as a medicine.[27, 28]

Ibogaine does have safety issues. It's known to be cardio-toxic, increasing blood pressure and leading to irregular heartbeat and even cardiac arrest. There have been around 30 reported deaths worldwide, most of them likely due to pre-existing heart disorders such as coronary sclerosis and cardiac arrhythmia. It has other negative side effects including nausea and vomiting as well as dizziness, psychosis and anxiety.[29]

Currently some addiction centers, in countries including Gabon, the site of origin of ibogaine, but also South Africa, Mexico, Thailand, Cambodia, the Philippines and Spain, offer ibogaine treatment for opiate dependence, in particular heroin. Some rehab programs using ibogaine are not well regulated; the quality of treatment and safety can vary. To minimize the cardiac risk, any treatment should be conducted in centers with medical oversight and cardiac monitoring.

Ibogaine has been approved as a medicine for treating heroin addiction in New Zealand. While it's known to help stop cravings and use of opiates, the exact mechanisms aren't understood, due to lack of research. It's thought to work on multiple neurotransmitter systems. We are planning to do brain imaging studies on ibogaine at Imperial College. For more on addiction, see Chapter 7.

## SALVIA
*Salvia divinorum*—the name translates as diviner's sage—is a member of the mint family, plants generally known for being calming, even soporific. This particular salvia, however, has been dubbed the "bungee jump for the brain."

### How is salvia taken?
It's an herb that's smoked or chewed. It can also be taken as a tincture.

### What does salvia feel like?
Feelings of detachment, sensory disturbances, hallucinations and dysphoria.

### How long does salvia last?
Five to thirty minutes when smoked, longer when chewed.

Historically, like psilocybin, salvia was used in rituals by Mazatec shamans in Oaxaca, Mexico. It produces visuals, hallucinations, tunnel vision, voices, physical impairment, dissociation and such profound alterations in consciousness that it makes you feel close to death. It is said to either put you in touch with the divine or be extremely unpleasant. In fact, so much so that most people take it only once. Those who do tend to be so-called psychonauts, committed to ticking off a list of every psychedelic experience.

Salvia is legal in most countries, although not in New Zealand, Australia, the UK or some U.S. states; it was listed as a contributing cause in a young person's suicide in the state of Delaware. It's since been the subject of a campaign to make it illegal throughout the U.S. This has not yet succeeded, possibly because, as mentioned above, it's very often a once-in-a-lifetime experience—or rather more accurately a "never again" experience![30]

The active ingredient is called salvinorin A and works on a type of opiate receptor called kappa. Opioid painkillers, such as morphine or heroin, work via another type of opiate receptor called mu. In 1988, I set up the psychopharmacology department in Bristol University with the pharmaceutical company Reckitt & Colman. We hoped to find a painkiller that worked via kappa rather than mu, one that might be less addictive than opioids. In animals, drugs that stimulate the kappa receptors can act as powerful painkillers.

We trialed a compound that we invented on a volunteer. Afterward, he told us he'd had a horrible time, had spent what felt like hours looking down onto his body from the ceiling, wondering if he'd ever get back into it. This is likely why no pharmaceutical company has ever developed a painkiller that targets kappa. We stopped the study.

In brain imaging—EEG and fMRI—the effects of salvia look very similar to those of ketamine or the classic psychedelics, according to a study from Johns Hopkins. This is interesting, as it's happening via kappa opiate receptors rather than serotonin receptors.[31]

However, people who take salvia do report distinct effects. And so there's an interesting mismatch between what imaging is telling you and subjective experience. This difference is currently under research at Imperial College.

The kappa receptors are in a different part of the brain than serotonin receptors, which is likely why the effects are subjectively different. This may also explain why salvia is generally unpleasant, whereas the classic psychedelics are not.

Salvia is unlikely to be used therapeutically in the future because people don't seem to come out of the experience with any insights. Although they do seem to be glad to be alive!

## AMANITA

*Amanita muscaria*, with its red or orange-red cap and white spots, is the mushroom of fairy tales, probably the most recognizable one in the world. Originally native to northern Europe and Siberia, it has now spread globally, traveling in the roots of exported trees.

### How is amanita taken?
The mushrooms are dried or taken as a tea. It also comes as a tincture.

### What does amanita feel like?
Depending on dosage, it makes you feel drowsy and relaxed, euphoric, have changes in perception and vivid dreams.

### How long does amanita last?
From four to eight hours.

Amanita's psychedelic ingredient is muscimol. This works on the GABA system of the brain, which is responsible for keeping it calm. The mushroom also contains a poison, ibotenic acid. When eaten by humans, ibotenic acid causes sweating, nausea, vomiting, diarrhea and seizures. Very rarely, people have died from eating amanita, mainly due to cardiac events. This is where it gets its nickname, fly agaric: ibotenic acid also kills flies and traditionally pieces of mushroom were sprinkled in milk to make an insecticide. When the mushroom is boiled in water to make tea, ibotenic acid is neutralized but muscimol isn't.[32]

At low doses, muscimol makes you sleepy and relaxed. It's still used in Siberia as a sleep tonic, as well as being rubbed on to joints for pain. With a large enough dose, muscimol causes hallucinations, in particular making things appear to change shape and size. This quality makes it very likely that amanita was the inspiration for the size-shifting effects of the "drink me" bottle and the "eat me" cake in *Alice's Adventures in Wonderland*, and possibly for the Super Mushroom in the game *Super Mario*, which makes Small Mario turn into Super Mario.

The change in perception is probably down to the fact that the part of the brain that's in charge of making decisions about size is loaded with GABA receptors. Muscimol can also get you stuck in thought loops, a kind of Groundhog Day of the mind.

There's evidence that amanita has been used by humans for thousands of years. In fact, there's a hypothesis that it was the origin of organized religion. *The Sacred Mushroom and the Cross: A Study of the Nature and Origins of Christianity Within the Fertility Cults of the Ancient Near East* by archaeologist Professor John Allegro, published in 1970, tells how early Christians used the mushroom to form communities and understand God. One of the pieces of evidence it cites is that a fresco in the thirteenth-century Chapel of Plaincourault in France shows Adam and Eve next to a tree of life that isn't growing apples but is one enormous amanita. At the time it was published, Professor Allegro suffered a loss of reputation for his controversial hypothesis, although it has since gained some support.[33]

Amanita may also be the source of the Santa Claus myth. The mushroom grows in northern Lapland, where people would often get snowed into their yurts by snowdrifts. Neighbors would be able to locate them via the smoke coming out of the snow from the chimney of their yurt. They'd then drop dried amanita down the chimney to keep up their spirits. If true, this would not only explain why Santa Claus wears red and white and comes down the chimney, but also why reindeer fly.

Because reindeer love to eat amanita, herders have been known to collect the mushrooms and scatter them to make the reindeer go where the herder wants them to go. Muscimol goes into the body and comes out of the body in the pee unchanged. When one reindeer pees it out, it has been known for others to drink their pee.

Apparently, people used to do this too. In the Middle Ages in Siberia, the mushrooms were claimed by the rich, who'd celebrate with amanita parties. Local serfs knew it could be recycled via urine so, when the partygoers came outside to pee in troughs, the serfs would collect the urine and drink it.[34]

\* \* \*

Amanita isn't illegal in most countries. This may be because the mushrooms haven't been as widely used in the West as have other psychedelics. And it may produce interesting experiences, but it almost certainly doesn't produce insights that make people question the establishment as the classic psychedelics do.

It's an interesting compound for research because at present we don't have any other medicines that work on the particular subtype of GABA receptors that it does.

A very similar molecule to muscimol, gaboxadol, was trialed as a sleeping medicine in the 2000s but wasn't licensed because of concerns it would be abused. Trials showed it had some risk of abuse, and the companies involved decided this outweighed its clinical benefit and stopped developing it as a medicine. In the past, synthetic muscimol has been trialed as a medicine for movement disorders, including Huntington's disease, but has never showed enough efficacy to be licensed.

Disclosure: I am a director on the board of Psyched Wellness, a company that has developed an amanita extract of muscimol for human use.

# Chapter 4

# WHAT HAPPENS IN YOUR BRAIN DURING A TRIP?

**WHAT GOES ON** in the brain during a trip remains one of the great unanswered questions in neuroscience, although a lot more is known now than 15 years ago. My psychedelic research group at Imperial College, led by Professor Robin Carhart-Harris, has done world-leading research in this area. Psychedelics are a powerful tool for giving insights into the brain basis of consciousness, and new imaging technology has been key to revealing those insights.

Professor Carhart-Harris came to my lab in Bristol as a PhD student over 15 years ago asking to do imaging studies of psychedelics. When we began this work in 2008, I assumed psychedelics would allow us to explore brain function and possibly give insight into the serotonin system of the brain and maybe even into the causes of depression. I didn't consider that they might become potential medicines in their own right.

# LOOKING BACK: THE SEROTONIN–DEPRESSION LINK

In the 1980s, part of my work was researching the serotonin system. You've likely heard of serotonin in relation to depression, perhaps as a "feel-good" chemical, or that antidepressant medicines such as Prozac are selective serotonin reuptake inhibitors (SSRIs). In fact, we still don't know if serotonin or the lack of it has a causative role in depression. Research has shown that depressed people have more serotonin receptors; one theory is that this upregulation is compensation for having a lack of serotonin.

At the time, there were older kinds of antidepressants available to treat depression, but this was well before SSRIs became commonplace. The final-line treatment for hard-to-treat depression was then (and is now) electroconvulsive therapy or ECT. In ECT, electricity is used to induce seizures in specific parts of the brain. The patient is given an anesthetic and a muscle relaxant, then an electric current is passed through the brain, in order to cause a seizure of 30 to 35 seconds. The person comes out of it a bit confused and disoriented, but their mood starts to lift within a few hours (more in Chapter 5).

ECT works because your brain is an electrochemical machine, made up of a web of around 100 billion neurons. Your machine's outputs—being awake, asleep, storing memories, swallowing and so on—are the results of millions of messages, flying around the brain web. Each message travels along the neuron via electricity but the connection that bridges the gap between neurons—the synapse—is a chemical messenger, a neurotransmitter such as serotonin. Neurotransmitters do this by acting on receptors. Some, like serotonin, act on multiple receptors.

We knew that ECT and the older kind of antidepressants affected the serotonin 2A receptors. And so, part of my research was finding out if blocking the 2A receptors might help with depression. We were working with a newly invented compound that did this, called ritanserin. In trials, we gave it to healthy volunteers and looked at their brains using EEG (electroencephalography). This looks like a swimming cap with wires sprouting out of it, and it reads brain waves via electrodes on the scalp.

Disappointingly, we found that after taking this 2A blocker it appeared that very little happened in the brain. We did notice one difference: the blocker gave people more of the large slow brain waves that occur during the deepest stages of sleep (which is why it's called slow-wave sleep).

We suggested 2A blockers might make a good sleeping pill, because deep sleep is restorative, good-quality sleep. But in trials, patients didn't report better sleep after taking them, perhaps because some felt "slowed-down" and woozy the morning after.[1]

When asked, they said they preferred existing sleeping pills (such as Nitrazepam), despite these not improving the quality of sleep. The result was that drug companies decided not to market 2A blockers.

I suggested the blockers might have another application: as an antidote to a bad trip. By blocking the 2A receptors, they stop psychedelics from working, which would make them useful in A&E. But this wasn't seen as a commercial target, so it was never developed. However, these blockers are now being used in studies to shorten LSD trips so people don't have to stay in hospital overnight.

After 2A blockers, the logical next step would have been to work on psychedelics themselves. But just as for most scientists at the time, the barriers to research—the stigma from working with an illegal drug, the cost of getting the substance and the difficulty of getting permission for the work—were too high. For these reasons, after the U.S. and UN ban on psychedelics from 1967 onward, clinical research fell to almost nothing.

That was the case until the 2000s. My team, first at Bristol University, then at Imperial College London, have set out to rectify this absurd—I would even say obscene—denial of scientific inquiry and potential therapeutic progress.

## A FIRST LOOK AT THE BRAIN ON PSYCHEDELICS

As expected, it took several years to get ethical approval and funding for our first study, to image the brain under psilocybin. The study finally started in 2008.[2]

We chose psilocybin for a number of reasons. It is very safe—there have been no proven deaths despite millions of people using it for thousands of years, across much of the world. In the UK, teenagers and twenty-somethings commonly make magic mushrooms into tea during the autumn growing season. Fresh mushrooms had only been made illegal recently, in 2005. Perhaps most importantly in order to obtain both university and ethical approval to do our study, psilocybin is not LSD. It doesn't have LSD's 50-year baggage of fear and misinformation.

We used volunteers who had previously tried psychedelic drugs, to minimize any risk of bad trips, particularly as they were being given a trip in a noisy, claustrophobic fMRI machine. This also helped with regulatory approval; we could state that we weren't introducing drugs to naive volunteers.

Because Schedule 1 drugs are governed by a mountain of regulations, their cost is ten times the usual cost of research drugs. To keep costs down, we had to administer the psilocybin via the IV route, as this reduces the dosage tenfold.

Another virtue of psilocybin is that it is relatively short-acting. This is especially true when it's given intravenously; the effects last thirty minutes, as opposed to four to five hours when it's taken orally. So even if someone did have a bad experience, we knew it wouldn't last long. In fact only one person of the many hundreds in our research has had a bad trip, despite many of them taking place in an fMRI brain scanner or while wired up to other machines. And that one bad trip was on LSD.

Conventional funding bodies refused to support studies on an illegal drug. We found financial and intellectual support from our collaborator Amanda Feilding and her charity, the Beckley Foundation, and had small donations from two psychedelic research bodies: MAPS (Multidisciplinary Association for Psychedelic Studies) and the Heffter Research Institute's associated charity. I donated money from my consulting and lecture fees.

Our funds still didn't cover the costs of the research, and certainly not the medical staffing and imaging analysis costs. The only way we were able to do the study at all was down to the generosity of the researchers themselves working for free on days off, weekends, evenings and nights, which I dubbed guerrilla pharmacology.

## TURN ON BECOMES TURN OFF

That first study looked at changes in blood flow in the brain using fMRI brain scanning as a proxy for how much brain activity is happening. One theory that had come out of the 1970s was that psychedelics increased blood flow in the brain. This seemed logical; research had shown that in delirious people, increased blood flow in the brain's visual system could be related to hallucinations, for example. But as nobody had done studies on drugs that cause hallucinations, we were breaking new ground.

First, we gave each participant a dose of saline, then scanned their brain for 30 minutes. This was to measure the activity of a normal brain. Then, we gave 2mg IV of psilocybin and scanned them again.

All the participants reported classic subjective effects, many describing them as strong. They reported simple hallucinations of bright geometric shapes, often rectangles or polygons, moving around the visual space, with an enhanced sense of color and brightness. People also described disturbances of visual stability with a sense of their visual field moving, as well as alterations in the perception of time and space. Less commonly, people reported spiritual and supernatural experiences. And a few described dissociating from their body and their self moving outside the scanner.

The self-reports were pretty much as expected. But then came the imaging results. And it was one of the most astonishing sets of results I have ever seen in a lifetime of science.

I had only once before had experiment results that were the exact opposite to what had been predicted. The first time, I'd coined a scientific maxim: If your results are opposite to what you predicted, they are very likely to be true!

We had predicted that we'd see an increase in brain blood flow, in particular in the brain's visual system. But the activity in the visual regions was unchanged. In fact, we didn't find any increases in activity, anywhere in the brain.

But all the subjects had hallucinations, so what was causing them? What we found was that in three regions of the brain, there was a profound *decrease* in activity. And the stronger the psychedelic experiences people reported, the more those three regions turned off. "Turn on, tune in, drop out!" had become "Turn *off*, tune in, drop out!"

This first unexpected finding kick-started a fascinating journey into the brain under psychedelics. At Imperial College, we are world leaders in this. We have completed three imaging studies on psilocybin. We have published the first paper defining the brain circuits of LSD.[3] And, more recently, we've conducted brain imaging into DMT[4] and are currently conducting studies with 5-MeO.

Working with Cardiff University, we've been able to use sophisticated, targeted technology to reveal brain activity in a more detailed and nuanced way, to look at its qualities as well as quantity. BOLD, an fMRI technique, measures brain activity more directly via changes in the level of oxygen in the blood. And MEG, a special type of EEG, gives a more precise measure of electrical activity in the brain. In the LSD study, because it lasts longer than psilocybin, we had time to do both BOLD and MEG on each subject on the same day. This had a great advantage; the two different but complementary imaging techniques could validate the other's results.

Each study we've done has brought us closer to understanding how psychedelics affect the brain. However, we are likely only at the very beginning of what this means for understanding consciousness. This chapter explains what we've learned so far.

## LOOKING INTO THE BRAIN

Your brain works as a series of connected networks. Each network works together on a specific function; for example there's one for vision, one for hearing and one for movement. There are also networks for attention, for planning, for evaluating the impact of the outside world, and one for making sense of it all.

The great challenge for the brain is integrating all these different networks, which each use billions of neurons. A special set of neurons—the Layer V pyramidal neurons—are in charge of doing this. The Layer V neurons perform at the highest level of our brain function. They also have the highest concentration of serotonin 2A receptors in the brain, the receptors on which psychedelics act.

Layer V neurons on an area of the brain called the anterior cingulate cortex (ACC) have the job of integrating motivation, emotions and memories. Those on the posterior cingulate cortex (PCC) integrate sensory inputs—seeing, hearing, positional sense and touch, etc. The communication network of the ACC and the PCC is the core of what is called the default mode network, or DMN (more on this later).

## WHAT HAPPENS IN YOUR BRAIN DURING A TRIP?

When you hear a noise, Layer V neurons coordinate the response. Your brain moves your head and body to optimize analysis of the sounds (PCC), checks your memory banks for previous exposure to the sound (ACC), then orchestrates an emotional as well as a cognitive response. Then, if the sounds are important, the brain lays down a new memory with all these different elements.

All this coordination happens via waves of electricity flowing between networks. These are brain waves, which can be measured via the skull, using EEG or MEG. Brain waves vary from the very slow ones of deep sleep to the very fast ones when you're awake and highly focused. When you sit quietly with your eyes closed, the most prominent one is called the alpha wave.

## WHAT HAPPENS WHEN YOU ADD A PSYCHEDELIC?

When someone takes a psychedelic, it slots into the 2A receptors in the Layer V neurons all across the brain. This activates and dramatically increases the excitability and firing of these neurons. In our studies, EEG and MEG showed this leading to a loss of the brain's typical and rhythmic pattern of brain waves. We saw the strong alpha waves breaking down and being replaced by shallower, unsynchronized waves.

This showed that the higher-level parts of the brain (the DMN), which usually integrate all the brain's input, had stopped communicating with one another. The drop in blood flow that we recorded in the first study was the DMN going offline.

You could compare the DMN to a conductor of an orchestra, keeping the instruments playing in synchrony. With the conductor gone, individual instruments will produce a series of unpredictable, discordant sounds. Instead of Bach, the orchestra plays free jazz. In the same way, with no central control, the different parts of the brain begin freestyling. The brain becomes chaotic and disorganized, the changed consciousness that we see under psychedelics.

## WHERE DO HALLUCINATIONS COME FROM?

Let's now look more deeply into the content of psychedelic experiences.

You may think of your brain and eyes as working like a camera, capturing what you see. But the truth is stranger. The brain does not video the world or take millions of pictures. This would quickly use up its memory capacity.

The brain allows us to "see" by creating a reconstruction of the outside world based on the signals that reach it from the retina. This process is explained in Figure 2.

As you look at something, photons of light from it enter the retina. Retinal neurons then send information as electrical impulses via the optic nerve into the visual cortex system.

One region of the visual cortex works out the location of the image, another works out if there is movement and in which direction, another works out the coloring and another the shape in 2D and in 3D. The visual cortex then pulls together all these analyses to create a construct of what's outside.

At the same time, higher brain regions are creating a prediction (which neuroscientists call a prior) of what you are seeing. Then the brain compares the sensory construct with the prediction and comes to a conclusion. If the prediction is wrong, it is then improved.

## Figure 2: How the brain sees

The brain creates an estimate of the outside world from interpreting electrical impulses from the retina. Under psychedelics the ability of the visual system to integrate these inputs is affected and so hallucinations emerge.

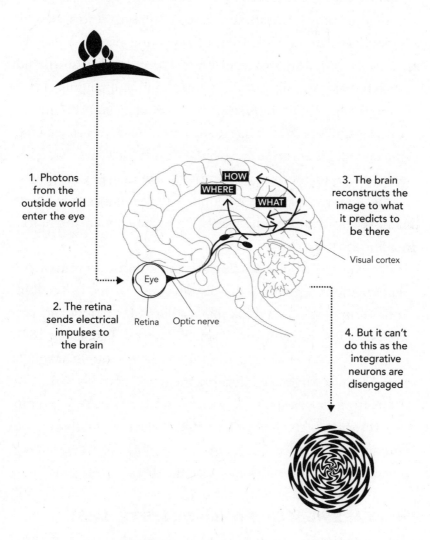

1. Photons from the outside world enter the eye

2. The retina sends electrical impulses to the brain

Retina    Optic nerve

Eye

HOW
WHERE
WHAT

3. The brain reconstructs the image to what it predicts to be there

Visual cortex

4. But it can't do this as the integrative neurons are disengaged

For example, you may have "seen" something that looked like appetizing food and tried to eat it. If you found it wasn't food but maybe a piece of colored paper, you learned this. The brain then reframes and improves the predictions so you don't repeat the mistake.

All your senses work like this, constructing your world by testing, then either validating or remaking priors. You have priors not only for what your brain expects to see, smell and touch but also for all your thinking, planning, imaginings and so on. Reality isn't "out there"; it's created by your brain.

The brain's software is analytical, continually reviewing and testing constructs in a similar way to modern AI. This makes the brain much more energy efficient than any computer. As Hermann von Helmholtz, one of the founders of modern neuroscience, said in the 1800s, "the brain is an inference-making machine."

For example, you're holding this book. How do you know? Sensory input tells you that you can touch it, see it, smell it. Your brain then says: I think that's a book. It's oblong, it's got words on, it's the right size, it smells and feels like paper. Your brain will keep analyzing the input that's coming through your eyes and fingers, confirming the book's bookishness.

Creating priors starts very early in life, as babies explore the world. It is the basis of learning. During childhood, we develop a lot of these priors in a short space of time. But we do keep developing new ones until death too.

## WHAT ARE WE SEEING WHEN WE SEE VISUALS?

Research in animals such as frogs shows that the very first stage of a brain creating images is the creation of simple geometric shapes. Usually, these are immediately fused into the whole and completed image that you perceive as the outside world. But under a psychedelic, the brain can't do this.

This explains why one common form of visual is pretty-colored patterns or shapes, often described as like Christmas tree lights. When someone sees this, they're seeing in the geometric shapes the early stages of the visual system in action. The last time this happened was in the first months of their life, before the visual system networks were fully developed.

Someone who'd taken magic mushrooms told me about his less common example of disrupted visual reconstruction. During a memorable, possibly unforgettable, trip, his whole world went upside down. To me this made perfect sense; the lens of the eye inverts the light from the outside world, so its image on the retina is upside down, but from a very early age our brain learns to flip this image. Probably, psilocybin interrupted this process, and so he was seeing the unflipped world that the retina sees.

## WHAT HAPPENS TO YOUR SELF UNDER PSYCHEDELICS?

Under psychedelics, not only do you see and experience things differently, you also think differently. This is because your brain not only creates predictions—the priors we have described—about the external world of your senses, but priors that rule your internal world too, for example, your language, cognition and emotions.

As the inputs for these priors are less concrete, they're harder for the brain to test and evaluate. It requires introspection, self-awareness and perhaps communication with others of our species. This work happens via a network of the brain called the default mode network, or DMN, mentioned before. (The PCC and ACC are the major parts of this.) The DMN's work is to orchestrate the content and focus of your mind, thoughts, plans and memories.

The DMN is the very highest level of the brain, the over-arching conductor of global brain function. It was identified around ten years before we started our imaging studies. The DMN is most active when we are relaxed and introspecting, for example, when we are contemplating ourselves,[5] planning the day ahead and reflecting on the past,[6] being rather than doing.[7] Babies have very little DMN connectivity, then it increases until adulthood.[8]

If you want to see the DMN light up in the normal brain, you need to ask someone to lie quietly in the scanner, eyes closed, not engaging in any outside tasks, just thinking about themselves. That's why it's called the "default mode" network —it dominates when all the other networks are off.

Under psychedelics, what we saw with both fMRI and MEG measures was the DMN going offline. Just as the visual system integration going offline leads to visuals, the DMN going offline explains many typical psychedelic experiences, such as wandering thoughts, altered thinking and vivid imagination.

## WHAT ELSE DOES THE DMN DO?

But the change in the DMN under psychedelics has more far-reaching effects too. It's the brain network responsible for your sense of self, the ongoing thinking process about you as a human being, including all your ideas and beliefs and attitudes.

That's why the DMN has also been called the self circuit. In Freudian terms, you might call it the source of the ego.

For some people, the priors of the DMN are mainly positive—happy memories, positive plans for the future, for example. But for those with mental health issues, for example, depression, OCD, anorexia or substance/behavioral dependences, they are not. The person may be over-engaged with a particular set of negative priors, such as feelings of guilt or low self-worth in depression; of compulsive cleanliness in OCD; or craving in addiction.

Even when priors aren't helpful, the DMN continues them. And because these priors are self-generated, it's hard for people to get outside their damaging thoughts or ultimately self-destructive behaviors.

Many people with dependencies, for example, don't want their thoughts and desires for alcohol or other drugs but they can't stop them, or they continue to use despite getting no pleasure from them. Using drugs, which started as a mind-made decision, ends up becoming a subconscious reflex behavior or habit that has escaped from conscious override. The behavior is now a prior, embedded in brain systems and below conscious control.

In the case of OCD, most people know that their compulsive thoughts or rituals are pointless but they can't stop them. Many depressed people feel the same about their negative thoughts but can't stop the habitual rumination.

There's a phrase coined by the eighteenth-century poet and painter William Blake that fits here: "mind-forg'd manacles." He used it about soldiers believing it was right to go to war, and then about workers choosing to leave the countryside to work in the "dark satanic mills" of the northern industrial towns. But it's also a powerful analogy for how the brain gets stuck in conditions such as depression and addiction.

Psychedelics may be a way to unmanacle the mind. When the DMN goes offline under a psychedelic, we think this gives a person a chance to step outside their habitual thoughts and beliefs about themselves, to escape their usual priors. For example, instead of thinking, "I am a worthless person," they might think: "I was abused as a child and so it makes sense that I feel worthless. And now that I know that, I can break away from it."

Some people in our neuroimaging studies reported an enduring positive mood after taking psychedelics. Some reported a profound inner peace; others felt happy and energized. As we will see in Chapter 5, upsetting the persistent but unwanted activity of the DMN may be a large factor in how psychedelics can help people get out of negative patterns, and so feel better.

**Making new connections**
Another unexpected discovery from our neuroimaging studies was the effects of psychedelics on connectivity in the brain.[9]

We worked with mathematicians to explore how the connectivity of the brain changes under psychedelics. Normally, each brain network tends to talk mainly within itself, less to other networks. This way of functioning is called the "small world" brain.

The small-world local connectivity relies on all the priors, for every situation, we develop from babyhood. It's very good at allowing us to function in daily life. And it's very energy efficient, which is why the brain is ten times more efficient than any known computer.

However, this efficiency comes at some cost, namely inflexibility and loss of creativity. And if your usual efficient connections that rule your self-worth and attitudes to life are maladaptive, it can lead to mental illness.

### Figure 3: Statistical diagrams of the level of connectivity in the brain

a) The normal brain. Most of the connections are made close to one another — the small-world brain.

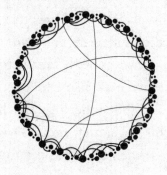

b) Under psychedelics there is much greater connectivity across the brain. This allows people to see and think about things with a very different perspective.

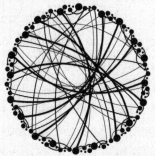

Under psilocybin and, our later studies showed, LSD, the connectivity of the brain changes to the picture at the bottom. As you can see, there is a massive increase in new connections between many more brain regions. The "small world" becomes a "large-world" network. Breaking down within-network communication allows much more between-network interaction. The brain is freed up, into a state it was last in during early childhood.

This helped explain the sense of mind expansion and universe engagement that people reported under psychedelics.

Our LSD study also showed that the greater the psychedelic experience that people reported, the greater the degree to which the usual brain networks broke up. The more complex hallucinations that a participant experienced, the greater the new connections between the visual cortex and the other parts of the brain. One example is synesthesia, seeing colors in sounds, which is the visual and auditory networks newly talking to each other.

Greater connectivity may also explain why people report having important insights under psychedelics, both personal and intellectual. This is a similar process to how you might be struck by new ideas when falling asleep, or after sleep, when the DMN also dampens down, allowing your brain to work without its top-down controller getting in the way.

New connections between brain regions that may have been isolated for decades may allow people to review old beliefs and issues, to relive memories of problems, to access hidden or repressed personal issues. It may be that the breakdown of top-down control by the DMN opens the mind, then these novel cross-brain connections allow the linking of memories and feelings with new insights, understandings and interpretations.

On top of this, psychedelics also produce enhanced neuroplasticity, which you'll read about in the next chapter. This allows new ideas and plans to be laid down in the brain, and endure well beyond the trip. This explains the long-lasting clinical outcomes of psychedelic therapy.

*    *    *

One of the most scientifically influential insights made under LSD must be that of the chemist Kary Mullis, who won a Nobel prize in 1993 for his method of decoding DNA. DNA is a coiled molecule, a double helix of strands of molecules. When Mullis took LSD, he saw snakes. Specifically, he saw snakelike portions of DNA unraveling and replicating.

He realized that if he could find a way of unraveling DNA in the test tube, and then of copying each unraveled section, he would be able to decode it much more rapidly. The answer was an enzyme, polymerase, which became his polymerase chain reaction (PCR). This is now a cornerstone of modern biology, allowing analysis of tiny samples, down to single cells of DNA, even from dead species such as mammoths, and enabling testing for anything from bacteria in wounds to horsemeat in our hamburgers.

That such a revolutionary discovery was made due to a hallucinatory state may not be as surprising as it at first seems. Albert Einstein said: "No problem can be solved from the same level of consciousness that created it."

## Opening your mind

William Blake wrote: "If the doors of perception were cleansed everything would appear to man as it is, infinite. For man has closed himself up, till he sees all things thro' narrow chinks of his cavern."

Blake's quote meant more than just "open your eyes." He understood that in all aspects of life, most of us fail to see the bigger picture.

In 1953, Aldous Huxley adopted Blake's words to name

his book, *The Doors of Perception*. He wrote that mescaline had opened up the chinks in his mind's cavern and allowed him to see things very differently. He deduced that if mescaline had opened his mind, something must have been closing it. And he decided that that something was the brain and so concluded, "the brain is an instrument for focusing the mind."

Huxley's hypothesis has been borne out by our imaging studies with psychedelics. We showed that psychedelics alter brain function in a way that disrupts the usual processing of external inputs and internal constructs. And this disruption changes consciousness, usually in a direction that is mind-opening.

In a way, psychedelics have also opened my mind. When we began our work, I didn't think that the brain images from our first psilocybin MRI studies would uncover a new, more powerful route to treat depression. That's the subject of the next chapter.

## HOW DO THE NON-CLASSICS WORK IN THE BRAIN?

We don't know so much about how all the non-classics affect the brain. It's often the same thing that's happening as with the classics—desynchronization—but more localized.

- Amanita (see Chapter 3) can produce alterations in the perceived size of objects. Its active ingredient, muscimol, stimulates GABA receptors in the visual cortex, altering function in the part that determines the perceived size of objects.

- Salvia works on kappa receptors in a brain network called the claustrum, which is about arousal and emotion rather than perception.
- Ketamine works on the glutamate system: it blocks a specific subtype of glutamate receptors called the NMDA receptors. As glutamate is the main "on" neurotransmitter for the brain, there are receptors for it all over the brain. And so while ketamine does work on the cortex like the classics, it also works elsewhere, including in the thalamus, the major relay station from the outside world into the brain. It also affects the hippocampus, the part of the brain that's like the RAM of memory. This might be why, although ketamine improves mood like classic psychedelics, the changes might not end up sticking.
- For MDMA, see Chapter 3.

## WHAT ELSE IS GOING ON IN A TRIP?

* **The self and the external world**

Another role of the PCC (posterior cingulate cortex) is giving a sense of being in a body that's somewhere in space. Under psychedelics, this self–external distinction breaks down. In a few participants, this breakdown was so great that they felt as if their bodies disintegrated, or they left their bodies and the scanner and went into space. For some, this felt like a spiritual experience. One described moving through space toward a bright distant light, finding himself bowing down at the feet of a deity.

* **Improved color vision**

  Psychedelics can enhance color vision, making colors brighter and more vivid. They can also sometimes allow color-blind people to see colors better. In our research studies, some participants with color blindness later reported this. Other color-blind people have written to me to share their wonder at a newfound appreciation of colors. We put a question into the Global Drug Survey asking color-blind respondents about the impact of psychedelic use—about half found that their color appreciation improved. Some scientists say this can't be possible, because color blindness is due to defective color cells (cones) in the retina. But the colors color-blind people do see can be enhanced by psychedelics in the same way as they are for people with normal color vision. My theory is that for most of us color isn't key, so our brains downplay its importance. Psychedelics disrupt this, allowing color appreciation to return to where it was when you were a baby.[10]

* **Music appreciation**

  Music may also open up more connections in the brain, which explains why it's a powerful means of enhancing the psychedelic experience and is now a key part of psychedelic therapy (see Chapter 6).

In fact, some researchers think that special music given in the right environment might have therapeutic properties on its own, i.e., without psychedelic drugs.

# Chapter 5

# CAN PSYCHEDELICS TREAT DEPRESSION?

*Psychedelics, used responsibly and with proper caution, would be for psychiatry what the microscope is for biology and medicine, or the telescope is for astronomy.*

Dr. Stanislav Grof

*Ian, 45, took part in the first study of psilocybin for depression at Imperial College in 2015. Here is his account.*

I have struggled with depression and anxiety since I can remember. Over time, the bad periods got worse, and the depression got deeper. Around ten years ago, I was working ridiculous hours, trying to do several different roles, and I burned out completely.

I tried talking therapies and six different antidepressants. At best, the pills took away my emotional range and numbed me in every part of my life. I went to see my GP yet again, and he said, "I don't know what you expect me to do for you." Which is possibly the worst thing to say to someone who's struggling in the pit of depression.

I tried things to help myself too, such as daily walks in nature—I just wanted to scream at the trees, "Why aren't I feeling better?" I tried journaling. Those were some of the most depressing journal entries you could ever hope to read.

I was desperate for that one thing that would make everything all right. At my lowest point, suicide seemed a logical choice. I thought it would be better not to be here, as it would stop me putting my wife and the people close to me through such hell. I can see the flaws in my reasoning now, but at the time it made sense.

Searching online one day, I found some videos of people who'd taken ayahuasca to help their depression. But by this point, it was hard enough for me to leave the house, let alone go to a South American rainforest. Then, I came across a lecture by Dr. Robin Carhart-Harris from Imperial College. He was talking about the effects of psilocybin in the brain, and he mentioned that Imperial were planning a trial into the effects of psilocybin on depression.

This felt like a last hope, although I tried not to frame it that way. And I didn't want to take a psychedelic. My friends and I had mucked around with psychedelics in my youth, and I had vowed never to take them again.

One night aged 20 or 21, I'd made the mistake of thinking the LSD had worn off, and so I took an extra half of a microdot. It was too much. I saw a light shining down on me, and a voice saying, "Come on Ian." I knew giving in to it would mean death or, at best, limbo. I felt my spirit leaving my body. My friends did their best to comfort me and look after me, but they were in the same altered state. Afterward I was scared to engage with the world, and it took me a few days to get back to normal. I felt I'd messed with something I couldn't trust.

But, two decades later, I signed up for the trial. A year later I was screened and, luckily, got through and into the trial. It was an open trial, so I knew I'd get an active dose, which was reassuring. Each person would be given first a medium dose of 10mg, to see if it suited them, and the 25mg therapeutic dose a week later.

During the 10mg dose my fear came up, and I really resisted the drug. I felt frozen on the bed. I was worried I would need the toilet, my way of clinging on to control. Afterward, I did feel a little lighter, definitely more carefree.

I was still extremely nervous of the 25mg dose. During that week, I met the guides and we spoke about my resistance. They were so helpful. They gave me full permission to be however I needed to be. They told me that if I needed to laugh, cry, shout, scream, roll around on the bed, it was OK. Whatever I needed to do physically and mentally, it was OK. They instilled in me that whatever came up, I could face it, using the saying "in and through."

Early on, during the first part of the 25mg dose, I remember going to the toilet with one of the guides waiting outside the door in the hospital ward, and I thought, "I don't ever want to feel like this again." I was still pushing against the drug.

The guides kept reminding me, "in and through." Then, an image of my abusive father came up in my mind, one I've had all my life. Normally, I'd have pushed it to one side. I thought if I ever engaged with the horrific memories around my dad, it would kill me or at least damage me beyond repair. The team helped me to go toward these demons, "in and through." As I did, it was like drawing back the curtain on the Wizard of Oz. I faced my father and saw he was not an omniscient figure who could crush me, but a pathetic little man.

Surviving that set the tone for the rest of the experience. I could face anything, deal with anything, nothing felt like a threat or a danger to me. I was hyper-present. I felt joy—I saw that the universe, every particle of nature, was in me, and I was connected to it. I felt compassion for every living thing, and then self-compassion for the very first time, because how could I exclude myself?

It wasn't all happiness and joy. At one point, the Jungle in Calais came to mind. I really felt the pain and suffering going on there in my heart. I was in floods of tears.

The biggest change in the session was that I approached everything that arose with openness, love and compassion. However uncomfortable or sad it was. Because I'd been able to face my dad, which I thought would break me, I could deal with whatever came up.

We cherry-pick the emotions we are allowed to feel, but I realized that every emotion that came up was welcome. I saw that crying is as valid a human expression as laughter.

In the follow-up session the day after, the team asked me if I felt depressed. I didn't want to jinx it, so it took me a while to respond. I said no, I don't feel depression, and I don't feel anxious.

The feeling of openness carried on after the session, for a good three months. This sounds so hippie, but I was open and welcoming to whatever emotion came up. This is the complete opposite to the depressed state, which is about doing all you can to avoid feeling the way you do. Normally, I'd have so much anxiety leading up to every human interaction, even just buying a loaf of bread in a shop. That quieted down to a huge extent. I wasn't constantly second-guessing myself. I didn't feel I was a new me, but that I was acceptable rather than having to bend and reshape to make myself acceptable.

Around three or four months later, some difficult life circumstances came up, and gradually that feeling of openness diminished. That is the hardest part of these treatments. For the majority of people, the effect does recede. I did find ways to reconnect to the experience, by listening to the playlists and diary entries I'd written during the trial. But slowly, the new way of living and being receded into being a mental concept.

Some things will never go back to how they were before. The experience helped me reframe my depression. Before, I saw it as a part of me I wanted to get rid of, a cancer I wanted to cut out. Now, I see it as an ongoing relationship I have with myself, that needs work, time and attention like any relationship. Again, it sounds hippie, but I can now approach those parts of myself with love, openness and compassion.

My depression is still a struggle. Last month, it was 8 or 9 out of 10. Now, it's a 7. As a participant in a trial, it's hard not to be able to access the treatment because it's illegal and not yet a medicine, especially when you've finally found something that works.

Together with another trial participant, I have set up PsyPAN, an advocacy network for trial participants. We aim to work with trial providers to make sure the participant is at the heart of everything they do, to help improve treatments and therefore the outcomes.

I do believe in the potential of these treatments. For many people they can bring a window of hope, relief and insight that —along with support and integration work—can help them to make changes that improve their quality of life. The "afterglow" does fade, so I hope that, in the future, people will be able to access the treatment, perhaps every three or six months.

More information at: Psychedelic Participant Advocacy Network (PsyPAN) psypanglobal.org.

Once psilocybin is available as a medicine, I believe it will be the biggest innovation in psychiatry treatment for 50 years. Severe depression shortens a person's life more even than some cancers; this can be due to stress-related illnesses such as heart attacks, because people stop eating and, of course, because they commit suicide.

I have worked as a clinical psychiatrist for nearly 50 years, and in that time I have treated a lot of people with chronic, severe depression. If they came to my specialist clinic, they would already have failed on several antidepressants.

The options for treating this kind of depression are currently limited. The first step is usually different combinations of antidepressant medicines, but up to 40 percent of people end up failing on antidepressants.[1]

That leaves two more invasive options. There's electroconvulsive therapy (ECT), explained in Chapter 4. If you've read or seen *One Flew over the Cuckoo's Nest* you'll think it's inhumane, but in fact it's not. With repeated doses—often twice a week for eight to ten weeks—it produces lasting changes in the brain and mood. And, in rare cases, there's focal ablative surgery (previously prefrontal leucotomy, colloquially called lobotomy), an operation to disconnect part of the brain, first done in the early 1950s.

However, these are serious and last-resort procedures. If psilocybin can replace them or, even better, stop people getting to the stage of needing them, that would be life-changing for patients—and for doctors too. There's a nice symmetry to our psychedelic trials; our new psychedelic research center has taken over the ECT floor of our hospital!

There is one psychedelic tool that's already in our depression toolbox: ketamine therapy. As it's legal, ketamine's antidepressant properties were discovered earlier than those of psychedelics, and it's already available as a medical preparation for anesthesia and for pain control. Later in the chapter, I go into the pros and cons of ketamine therapy for depression. It can be an excellent therapy for some people (see page 110).

## TAKE A TRIP TO BETTER MOOD

In 2008, when we began our first psilocybin imaging study, there were two recent small studies suggesting that taking just a single psilocybin trip might leave people feeling better about life. One came from Professor Roland Griffiths's group at

Johns Hopkins medical school. They gave nondepressed people quite a high dose of psilocybin (30mg orally) and therapy, using the stimulant methylphenidate (Ritalin) as an active placebo control. The psilocybin group had a significant improvement in mood and well-being, but the Ritalin group did not. Most rated the experience as among the five most personally meaningful and the five most spiritually significant experiences of their lives.[2]

The second study, by Professor Charles Grob at New York University, was in people with advanced-stage cancer who, not surprisingly, had anxiety and depression. It also showed improvements in mood, even six months after the study ended.[3]

Those results encouraged us to ask the participants in our psilocybin brain-scanning studies how they felt in the period following treatment. Our subjects were not depressed people, but some had powerful positive mood shifts that lasted for some weeks. One said:

Ever since (the scan) I have found it much easier to engage in the moment—to attend to the here and now; whether this be watching water in a fountain or sitting in science talks and meetings this morning. There were some fountains in Cardiff and the water was being blown by the wind, allowing the sun to highlight the spray. I could have watched it for ages—somehow the beauty was enhanced . . . Whatever it was, it has lasted—like the sun shining through the leaves this morning—it made me slow down my walk to work and enjoy the experience of it flickering over my face.[4]

## WHAT DOES PSILOCYBIN DO IN THE BRAIN?

How could psilocybin, a drug that lasts for four or so hours, make people feel better or improve their mood for days, or even weeks or months?

We had two clues, namely two areas of the brain that our first imaging study showed were changed under psychedelics.

The first was a small area of the brain called the CG25, part of the ACC that we mentioned in Chapter 4. Psilocybin produced a clear reduction in activity in CG25.

Around ten years previously, Dr. Helen Mayberg at Emory University in Atlanta, a pioneer in mapping the brain circuits involved in depression, had turned the field of depression on its head. Previously, it had been thought that depression was due to a failure of a mood-maintaining brain circuit, but Mayberg had the revolutionary idea that it was down to a depression-maintaining circuit, driven by overactivity in CG25. And she proved this with a very clever experiment that turned off CG25.[5]

Mayberg adapted a type of neurosurgery that had been developed for Parkinson's, deep brain stimulation (DBS), where a neurosurgeon implants electrodes in a particular part of the brain. For Parkinson's it's in the movement circuit, but in Mayberg's experiment it was CG25. When the electrodes are turned on, the brain area is switched off.

For this kind of operation the patient must be awake on the table, so the surgeon can check they've got the electrodes in the right place. In Mayberg's experiment, when the electrodes were turned on and CG25 was switched off, some patients reported an almost immediate lifting of their depression. Some said it happened more slowly. Later trials of DBS showed that although it doesn't work for everyone, in some patients the results can be life-changing.

Research has shown that other ways of treating depression also reduce activity in CG25. These include ketamine (more on this later), other antidepressant drugs, cognitive behavioral psychotherapy (CBT), and even (sometimes) a placebo. We knew psilocybin turned down CG25: the question was, would it be antidepressant?

The second key area of the brain was the default mode network (DMN), aka the self circuit.

Before our study, brain imaging research had shown that in patients with depression the DMN is overactive and over-connected. And the *greater* the extent of connectivity in the DMN of depressed patients, the greater the degree to which they ruminate on their negative thoughts.[6]

The thoughts and memories of depressed people run along rails of negativity and guilt from which they can't escape, called "tramline thinking." This negative thinking develops a life of its own, pushing the person deeper into their depressive ruminations.

You can get a sense of what this feels like from this description by Scottish philosopher Thomas Carlyle: "It was one huge, dead, immeasurable steam-engine, rolling on, in its dead indifference, to grind me limb from limb."

As we had shown in our first study that psilocybin disrupts the DMN, we thought it might break the ruminative process and, in that way, lift depression.

## IMAGING TRIAL 2: THE FIRST MODERN
## PSILOCYBIN IN DEPRESSION RESEARCH[7]

We decided to study patients with the greatest need, those with treatment-resistant depression.

We knew it wouldn't be easy to get the necessary permissions for a psychedelic trial on people with depression, the first one in the UK for 50 years. In the end, the whole process took 32 months. Raising money was the easiest bit. The Medical Research Council, the main UK funder for experimental brain research, agreed to fund a small double-blind controlled study as part of a program to find new treatments for resistant depression.

However, it took a full year to get ethical approval. I had to go in front of the local hospital ethics committee three times. There was one GP but no psychiatrists on the committee, nobody with any firsthand experience of treating people with severe treatment-resistant depression. As well as me, our delegation included two professors of psychology, Professors Valerie Curran and Steve Pilling. The committee kept insisting it was too dangerous to give psilocybin to depressed people. One of our arguments was that in the 1960s it was a medicine in some countries—under the name Indocybin—that was used for depression and anxiety. And also that probably a million people a year in the UK take magic mushrooms with no adverse effects—and some of them must be depressed.

Eventually the committee gave permission, but only for a preliminary safety trial in 12 patients. It had to be open label—meaning that both patients and doctors knew they were getting psilocybin—with a six-month checkup.

It then took a further 12 months to get hold of the drug itself, because of all the international rules around it being a Schedule 1 drug.

To be on the trial, the patients needed to have failed on two standard antidepressant drug treatments. One had tried 11 different drugs. All but one had tried and failed to respond to cognitive behavioral therapy. Some had been depressed for over 20 years.

The treatment sessions took place in a quiet room that we made more homely and welcoming than the usual sterile and impersonal medical outpatient clinic space. It had subdued lighting, wall hangings and pleasant music. There were two psychiatrists or psychologists on hand to guide the patients through the experience.

First, each patient had an fMRI scan. Then we gave each one a 10mg dose by mouth to make sure they'd tolerate the 25mg treatment dose. All did, so a week later all 12 received the full 25mg dose. Most of, but not all, our patients had a strong response to this, lasting around four hours.

The day after the treatment, the subjects had a second fMRI scan. They then talked through their psilocybin experience with their therapists. This step is *integration*, a key part of the therapy for patients to maximize the insights and therapeutic processes the trip initiates (more on this in Chapter 6).

Although the primary outcome was to test the safety of a depressed person taking psilocybin, we also measured mood using standardized rating scales. At one day, one week and two weeks after the 25mg dose, almost all scored much lower on scales of depression. At two weeks, 10 out of 12 patients met the criteria for recovery from their depression, a remarkable outcome. As the results were so encouraging and there were no serious adverse effects, the ethics committee granted us permission to continue the study with another eight subjects. Most of the extra eight had a similar large reduction of symptoms.

The results showed a single 25mg dose of psilocybin plus therapy produced a more powerful antidepressant effect in these difficult-to-treat patients than any other single-dose treatment available. Within a week—and often within a day—it halved depression scores. For a few of the patients in our study, their depression didn't come back for over eight years. But sadly, for most, their depression began to creep back over the next six months. Similar results are found with all treatments for severe depression; when they are stopped the depression tends to come back. How to help prevent this recurrence is one of the key research questions in the medical psychedelics field today. Even so, we now know that for most people there is a good chance psilocybin will lift treatment-resistant depression, giving patients hope. This is the reason the Australian government has now approved it for this indication.

## THE PSYCHEDELIC GOLD RUSH

Shortly after we published, two U.S. studies confirmed the antidepressant effects of psilocybin-assisted therapy. The first was from the team led by Professor Roland Griffiths at Johns Hopkins[8] and the second from Professor Stephen Ross at New York University,[9] and they were both in people with life-threatening cancer. The NYU study said, "psilocybin produced immediate, substantial, and sustained improvements in anxiety and depression and led to decreases in cancer-related demoralization and hopelessness, improved spiritual wellbeing, and increased quality of life." And this lasted more than six months.

Not surprisingly, several small pharmaceutical companies started to develop psilocybin therapies. The one that's furthest along, Compass Pathways, has recently completed a randomized, double-blind study of COMP360 (its proprietary synthetic psilocybin) for treatment-resistant depression at 22 sites across Europe and North America. It tested doses of 25mg and 10mg, using 1mg as a placebo. Over a quarter of patients on the 25mg dose were in remission at 12 weeks.[10, 11]

## ESCITALOPRAM STUDY

For our next study we compared psilocybin with the gold standard SSRI treatment, escitalopram. We wanted to see which one worked best in patients, and also to use fMRI imaging to explore our theory that they work differently in the brain.[12]

It was a double-blind study; patients were randomized either to a psilocybin treatment or to escitalopram for 6 weeks. We recruited 59 people with moderate-to-severe depression. They were divided into two groups. The psilocybin group had two 25mg doses of psilocybin, one at the start, then one three weeks later, plus six weeks of a daily placebo in place of the escitalopram. The escitalopram group also got psilocybin and at the same times, but at a 1mg (placebo) dose plus six weeks of escitalopram (10mg for three weeks, then 20mg for three weeks). All patients had integration and psychological support for each psilocybin session.

Results showed psilocybin worked faster and better than escitalopram on most outcome measures. At six weeks, it worked twice as well at getting people into remission, 57 percent of the psychedelic group versus 28 percent of the antidepressant group.

## Figure 4: Impact of psilocybin and escitalopram on depression scores

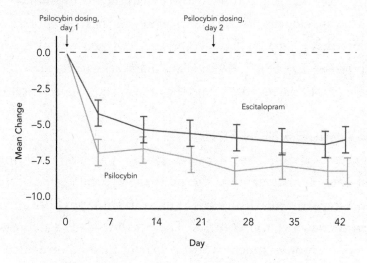

Source: Carhart-Harris R, Giribaldi B, Watts R, Baker-Jones M, Murphy-Beiner A, Murphy R, Martell J, Blemings A, Erritzoe D, Nutt DJ. et al., 2021, Trial of Psilocybin Versus Escitalopram for Depression, *New England Journal of Medicine* 384: 1402–1411, ISSN: 0028-4793.

## SSRIs VERSUS MUSHROOMS IN THE BRAIN

So we'd shown psilocybin was working on depression, but more importantly we wanted to know *how* it was working. We used the fMRI imaging from the two studies to ask these key questions: What is psilocybin doing in the brain to alleviate depression? And does it work differently from other antidepressants? The results were, frankly, thrilling.

We already knew how escitalopram, in the same way as other SSRIs, lifts mood in the brain. It works via the emotional circuits, specifically on the amygdala, the part of the brain that acts like an alarm system, responding powerfully if people are stressed or fearful. In depressed people, the amygdala fear response is supersensitive. This sensitivity drives low mood but also other depression symptoms such as loss of appetite and insomnia. This may explain why in Figure 5 the well-being scores for escitalopram were lower than those for psilocybin.

Our imaging results from both depression trials supported our theory that psilocybin has a completely different way of treating depression than SSRIs. We did two kinds of imaging—one looking at emotional responses and the other brain connectivity.

## The emotional effect of SSRIs

Before treatment, we looked at the patients' emotional response in the scanner. We used the typical test: showing them negative (fearful) faces and positive (happy) ones, with neutral ones as the control. The amygdala of a depressed person responds much more to fearful-face pictures than a nondepressed person's does.

After six weeks we saw, as expected, the blunting of the amygdala response in the SSRI patients. This is the start of the antidepressant effect. By reducing emotional distress, the SSRI allows the person to respond more normally in situations they would have previously found stressful and/or threatening. They become more engaged in life and with people and so their mood lifts.

That's why SSRIs can take six to ten weeks to work. They protect the emotional brain from stress, allowing it to heal. In this way, SSRIs are like a plaster cast that you use to set a broken limb. The plaster doesn't heal the bone but it protects it from stress so it can heal naturally. If you keep taking them, SSRIs protect against further episodes of depression.

However, there's a drawback to suppressing the emotional circuit, and it's one that a lot of patients complain of. Along with suppressing negative emotions, SSRIs also suppress positive ones. This may explain why in Figure 5 the well-being scores for escitalopram were lower than those for psilocybin.

Figure 5: Impact of psilocybin and escitalopram on well-being scores rated on the Warwick Edinburgh Mental Wellbeing scale

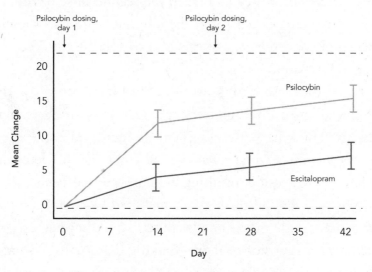

Source: Carhart-Harris R, Giribaldi B, Watts R, Baker-Jones M, Murphy-Beiner A, Murphy R, Martell J, Blemings A, Erritzoe D, Nutt DJ et al., 2021, Trial of Psilocybin Versus Escitalopram for Depression, *New England Journal of Medicine* 384: 1402–1411, ISSN: 0028-4793.

Looking at psilocybin in the scanner, we could see it had no effect on the emotional system, which tells us that it lifts depression via other mechanisms.

## PSILOCYBIN AND MAKING CONNECTIONS

In our first imaging study, done in healthy subjects, we saw decreased DMN activity leading to increased connectivity in the brain. We saw this in the first depression study too, despite the fact we scanned the patients the day after the trip, when all of the drug had gone from the brain. (We didn't think it was safe or ethical to give depressed patients a trip in the claustrophobic environment of the scanner.) Remarkably, the amount of connectivity on day two predicted the patients' depression improvements at six months.

In this, the second depression study, we imaged the patients three weeks after they had taken the second dose of psilocybin. Again, to our surprise, their brains were still showing increased connectivity, and this also correlated with recovery from depression.

Based on these results, we think instant brain disruption produced by psilocybin frees the person from their rigid, negative thinking patterns. Then the increased connectivity allows them to find new ways to overcome the problems that led to the depression, resulting in improved well-being.

What we saw on the scanner mirrored what patients told us. Most said that their minds felt freer and they had escaped the internal loops of negative thinking that are the hallmark of depressive thought processes. Along with more flexibility of thinking, they described greater connectedness to other people, to nature and to the world. They often used a computer analogy to explain the experience, such as "reformatting" or "defragging" or "rebooting" a computer hard drive to make it work smoothly:

I felt my brain was rebooted; I had the mental agility to overcome problems. It was like when you defrag the hard drive on your computer. I experienced blocks going into place, things being rearranged in my mind, I visualized as it was all put into order, a beautiful experience with these gold blocks going into black drawers that would illuminate and I thought my brain is bring defragged, how brilliant is that!

The reset switch had been pressed so everything could run properly, thoughts could run more freely, all these networks could work again. It unlocked certain parts that were restricted before.

# WHY DO SOME PEOPLE GET WELL, BUT OTHERS DON'T?

Three months after the first depression study, about 50 percent of the patients had kept their improvements, and half of those were in remission, i.e., 25 percent were no longer depressed. But the other 50 percent had slipped back toward low mood. One was worse than before treatment but only very slightly so. In the second depression study, again a quarter were in remission at three months.

As yet, we don't know why some people get well and some stay that way, but in others the depression creeps back. There's a lot of interest in the factors that predict response, but we don't yet have enough data. There are some pointers: ·

## INSIGHT AND MASTERY
In our studies, those who did well seemed to get insight into the causes of their depression and to achieve some mastery over these.

One patient said:

My outlook has changed significantly. I'm more aware now that it's pointless to get wrapped up in endless negativity. I feel as if I've seen a much clearer picture.

Another said:

My mind works differently [now]. I ruminate much less, and my thoughts feel ordered, contextualized. Rumination was like thoughts out of context, out of time; now my thoughts feel like they make sense, with context and logical flow.

## A MYSTICAL OR RELIGIOUS EXPERIENCE

In our study, we measured the size of each patient's positive psychological experience under the drug. We called this the "peak" experience to avoid any religious connotations, though for some this experience will be religious. For others, it will be more mystical or spiritual.

We used a scale designed for this purpose—the Altered States of Consciousness Scale. We found that the more patients scored on this, the better their antidepressant response. The two people who had mild trips didn't seem to do as well. There's more on the importance of mystical experiences in Chapter 9.

## ANXIETY DURING THE TRIP

Those who felt more anxious during the treatment tended not to get a good outcome, a finding the Johns Hopkins group had made previously too.

## DEEP-SEATED AND LONG-LASTING DEPRESSION

It is perhaps not surprising that some people's depression doesn't disappear with a single psilocybin treatment if it has been going on for their whole life.

For many people, depression starts with abuse in childhood, which causes trauma. We know that children tend to believe that they have more effect on the world than they really do. So when things go wrong, even if this is due to abuse from parents or other adults, they tend to think it's their fault, to blame themselves. This early form of depressive thinking becomes the default of their emotional life as adults. This makes breaking free very difficult. Psilocybin treatment may disrupt this and allow a period of normal mood, but the person's underlying brain processes that produce depressive thinking will gradually take over again.

## REAL LIFE GETTING IN THE WAY

If the factors that made the person depressed in the first place haven't gone away, they're more likely to slip back into depression. Many studies have shown that ongoing life stresses are one of the strongest risk factors for depression relapse. Traditional antidepressants such as escitalopram work at a chemical level to protect against stress pushing the brain back to its depressed mode—so long as they are in the brain. This is one way to produce resilience to depression. Psilocybin works in just one dose to alter thinking processes and improve well-being, which is another way to enhance resilience.

# DO PSYCHEDELICS GROW YOUR BRAIN?

One reason for the behavioral effects of the classic psyche-delics, and the fact that they last, is that they stimulate neuro-plasticity. Animal studies suggest this is the case for ketamine and MDMA too. Neuroplasticity is the ability of the nervous system to change by forming new neural connections, to rewire itself. It's how you learn and remember, as well as adapt to whatever life brings. Children's immature brains are always doing this on a larger scale, but, as we go into adulthood, we have less neuroplasticity and so we get stuck in our ways of thinking.

## WHAT DOES DEPRESSION DO TO BRAIN CELLS?

We have known for years that neuroplasticity is disrupted in people with mood disorders and addictions.[13]

Even before fMRI existed, CT scans showed that depressed people have brains that are smaller than nondepressed brains. In fact, middle-aged people with chronic depression have brains the size of someone ten years older. You need a hormone called brain-derived neurotrophic factor (BDNF) to keep neurons functioning and to promote neuroplasticity and we know that stress reduces levels of this.

For 20 years, we've also known that antidepressants slowly restore levels of BDNF, helping neurons grow more normally and restoring the normal architecture of the brain. For exam-ple, after six weeks of taking escitalopram, patients are close to normal neuronal function.[14]

## WHAT DO PSYCHEDELICS DO TO BRAIN CELLS?

Psychedelics trigger neuroplasticity immediately, which is why they've been dubbed psychoplastogens. It's not yet known, though, how long this plasticity lasts. In our studies, we measured neuroplasticity of the visual system after the trip and found there was a bigger response in electrical activity, a sign of a plasticity boost, up to a month after the trip. But some people in the studies describe feeling different for months and even years after the drug has left the body, so some neuroplasticity must last long enough to establish new habits and thought patterns. It may be that some people will require top-up doses every few months or years to maintain a state of non-depression.

## HOW DO PSYCHEDELICS PRODUCE PLASTICITY?

Once we'd published our first imaging study, other scientists became interested in looking at what psychedelics were doing at the level of brain cells, using both test-tube and animal studies.[15]

It's been now shown that psychedelics increase levels of BDNF, even while the trip is happening. And they do grow the brain too. Each brain cell (neuron) has many hundreds of branches (dendrites), each of which has thousands of small spines (synapses), and psychedelics increase the number of both branches and spines.

There are several companies and scientific groups developing neuroplastic drugs without the psychedelic effects, which they've named non-hallucinogenic plastogens. The U.S. Department of Defense has put $26.9 million into Professor Brian Roth's group, who are researching this in the U.S.[16]

Figure 6: Schematic of neuroplasticity showing how a single dose of psychedelics can increase the number of dendritic branches on a neuron

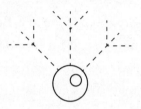

Before psychedelic

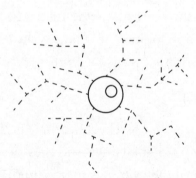

After psychedelic

Plasticity in action: neurons have an increased number of dendrites and synaptic spines after taking psychedelics.

## WHY DO WE WANT A MORE PLASTIC BRAIN?

If the trip itself gives you insights and new perspectives and takes you out of your usual thought grooves, plasticity along with therapy will help you develop new, better patterns of thinking. If people feel more positive after a trip, the plasticity can help enhance that by training the brain in its new positive grooves.

## DO OTHER CLASSIC PSYCHEDELICS TREAT DEPRESSION?

There have been studies into ayahuasca in retreat settings that show promising results for depression.[17]

It's also being developed as a treatment in South America, where the first study has provided evidence of efficacy in depression from a single psychedelic dose. This trial was conducted using a placebo-controlled design, and it showed antidepressant effects the day after taking ayahuasca that lasted for at least a week.[18]

At Imperial College we are doing studies into DMT, ayahuasca's active ingredient, which have shown a very similar brain effect to psilocybin and LSD.[19, 20]

The pharmaceutical company Small Pharma (mentioned earlier) has trialed an infusion of DMT with a trip lasting 20 minutes, to see if it can produce enough effect in the brain to lift depression. It could be useful to have a shorter-acting treatment than psilocybin. The results of their first clinical study are very promising.[21]

There have also been studies on 5-MeO, another short-acting substance, on people with treatment-resistant depression.[22]

# CAN PATIENTS HAVE PSYCHEDELIC THERAPY IF THEY'RE TAKING ANTIDEPRESSANTS?

Most trials ask people to stop taking SSRIs a few days or weeks before starting. In general, if someone is on an SSRI antidepressant, it is safe to take psychedelics but the effects will be diminished. People on SSRIs tend not to get the emotional release and peak experience. In the Compass Pathways study (discussed earlier) there were 20 people who were on SSRIs and they reportedly got good effects. Also, a Swiss study gave psilocybin to healthy volunteers who'd been taking escitalopram for two weeks and found that they still experienced psychedelic effects. Professor Matthias Liechti's group have also done some pilot work in patients who are taking SSRIs that suggest that a high dose of LSD might work.[23]

Some older antidepressants, such as amitriptyline and mirtazapine, block the effects of psychedelics altogether as they block the 2A receptor. The same is true of some antipsychotics such as risperidone and olanzapine. Patients need to come off these before taking psychedelics. However, patients can have ketamine therapy and keep taking SSRIs; esketamine is licensed as an add-on to SSRIs.

## CAN KETAMINE TREAT DEPRESSION?

Ketamine therapy is available now, and it's something of a revolution; it is the first pharmacologically novel antidepressant in over 30 years. Its effect on mood was discovered in the 1990s by Professor John Krystal, a psychiatrist and addiction specialist at Yale.[24, 25] At the time, Krystal was using ketamine in his experiments to mimic some symptoms of schizophrenia, and he noted that test subjects were reporting a mood lift for a day or so after the experiment. (It's interesting that, a few years later, our group discovered the mood effects of psilocybin by a similar process.)

Krystal and his team started studying this effect. A trial in patients with resistant depression found ketamine had an immediate mood-lifting effect, even as patients came out of the dissociation and confusion that the drug produces. The next day they still felt better, but the effect gradually wore off; by the time a week had passed, they'd slipped back into their previous depression.

Since this pilot trial, over a dozen studies have confirmed ketamine's usefulness in treatment-resistant depression.

## WHAT DOES KETAMINE DO IN THE BRAIN?

Ketamine works by blocking the normal function of glutamate, the brain's primary neurotransmitter. The cortex, the high-level thinking part of your brain, is packed with glutamate receptors. So it's not surprising that ketamine totally distorts and disrupts the brain's usual function, allowing the breakdown of patterns of depressive thinking.

We did our ketamine imaging study after we did the psilocybin study. Because of the wide action of glutamate, the ketamine results were not unexpected, but, as you've read, the psilocybin results were.[26]

Like the classic psychedelics, ketamine also leads to brain plasticity, although we suspect it doesn't last as long.[27]

We are now using a new imaging tracer drug, UCB-J, that measures changes in the number of synapses to investigate this, comparing the amount of plasticity from ketamine with that of DMT.[28]

With both disrupting and plasticity effects, ketamine is suitable for treating depression and also drug and alcohol dependence (for more, see Chapter 7) and possibly also for behavioral addictions such as gambling and pornography.

## HOW IS KETAMINE TAKEN?

Ketamine is usually given either by intravenous or intramuscular injection. It's an off-label treatment for depression because it is currently licensed as an anesthetic. This means it must be prescribed on a named-patient basis by a doctor. Some clinics offer ketamine only, and some offer ketamine-assisted psychotherapy. The psychiatrist Professor Rupert McShane in Oxford, England, has run a ketamine clinic for many years, using it in place of ECT for patients who are resistant to prior treatments. The treatment is a course of four to eight IV ketamine injections given by an anesthetist over two to four weeks.[29]

At the Awakn group of clinics, for which I am chief research officer, ketamine is given as an intramuscular injection along-side psychotherapy. We give one dose a week for four weeks. We find that 30 percent of people with depression dramatically improve, about 40 percent say it was worth coming and 30 percent say it hasn't helped.[30] We think the psychotherapy we use helps the results last longer.

The newer nasal spray form, esketamine (Spravato), is being marketed as a licensed treatment for depression that has not adequately responded to SSRIs by the pharmaceutical company Janssen. It's usually two treatments a week for three to four weeks, then one a week until a full response. It's expensive at around $10,000 per course of treatment but still is supported by many health insurers.[31]

Ketamine also comes in an even newer form: Professor Paul Glue and his team at the University of Otago in New Zealand have developed a controlled release oral tablet, which is currently being trialed.[32]

# THE UPSIDES AND DOWNSIDES
# OF KETAMINE THERAPY

## Upsides

* It's available now in the UK and U.S. as well as many other countries. Even though it's a controlled drug, it's also a licensed medicine. That means that, unlike psychedelics, a doctor can prescribe it off-label.

* If you have treatment-resistant depression, it gives you another treatment option.

* Unlike psychedelics, we know ketamine works if you are on SSRIs, because it acts on glutamate rather than serotonin receptors. In fact, the license for esketamine states that it is to be taken on top of SSRIs, for people who have only a partial response to SSRIs.

## Downsides

* Some people do well on it, some don't. McShane's figures show that about a third do well after a few doses, a third get some benefit, and a third don't benefit at all.[33]

* It needs to be repeated. The effect of one dose is short-lived, around two to three days when administered IV/i.m. and without psychotherapy.

* It's not cheap. Some U.S. insurance companies won't cover off-label use. The intravenous version needs an anesthetist present, which puts up the cost too.

* Some people don't like the feelings that come with dissociation: bizarre images, feelings of going mad,

being powerless and unable to move, plus feelings of body shape and size distortion. Attempts are now being made to find a similar chemical that lifts mood without these effects. Some people do like these feelings, which leads us to the next downside: you can become addicted to ketamine.

* You can develop both tolerance and dependence to ketamine. And taking high doses regularly has two serious health effects: bladder and brain damage. However, these harms are much more likely with recreational use. For more detail on harm, see Chapters 12 and 13.

## DOES PSYCHEDELIC THERAPY WORK FOR PEOPLE WHO HAVE ANXIETY?

In our first depression study, confirming earlier work, anxiety during the trip predicted a poorer outcome. There are groups interested in trialing psychedelics alongside anxiety medication, such as a benzodiazepine or an SSRI, to see if it can improve outcomes. N.B.: As we showed in the escitalopram study, prior treatment with SSRIs may reduce the power of the psychedelic.[34] (For more on psychedelics as a possible treatment for anxiety, see Chapter 10.)

## HOW GOOD IS THE EVIDENCE ON PSILOCYBIN AND DEPRESSION?

Our early trials were "open," which means that the subject and the researcher both knew that the person was getting the drug. Open trials almost always get better results.

**DOES PSILOCYBIN WORK FOR NORMAL DEPRESSION, NOT JUST TREATMENT-RESISTANT?**
It hasn't yet been trialed specifically for that, but in our psilocybin-escitalopram trial, some patients who were not treatment-resistant responded well. There is new real-world evidence coming out on DMT and depression from the Small Pharma study, mentioned on page 109.

Our trial that compared psilocybin with escitalopram was double-blind; nobody knew who was getting what. But the effects of psychedelics are so distinctive that it's likely everyone who had the psilocybin knew they'd had it, as did the therapists.

That said, the evidence is stacking up in favor of psyche-delics. In a 2021 study on depression with an interesting "wait-ing list" design, half the group started with two doses of psil-ocybin alongside supportive psychotherapy. The other half waited for two months, then had the same treatment. A year later, 58 percent were in remission.[35]

In our clinical trial, over half went into remission too. However, there are still a lot of questions that need to be researched. We don't yet know the best way to treat people. We don't know enough about how psychedelics work with SSRIs. We don't know about optimum dosages or regimes. And we don't know how to make sure the effects last. But I am very hopeful that the promis-ing results of the early studies will translate into an effective treatment so that, in a few years, psychedelics will become an important alternative treatment for depression.

## Chapter 6

# THE HISTORY AND PRACTICE OF PSYCHEDELIC THERAPY

**IMAGINE YOU'RE AT** a work party, and you find out that the glass of punch you just drank contains a massive dose of LSD. In 1954 this really happened—at, of all places, the CIA's annual Christmas party. How could something that sounds so improbable be true? Because during the 1950s, CIA operatives carried out a series of in-house experiments to find out what happens when people are given LSD without warning. The protocol was to tell people after they'd taken the drug, but before the effects started.

The previous year, Frank Olson, an army scientist, had been spiked with LSD at a work retreat. At first he was quiet and withdrawn, then he went into a deep depression. He became paranoid, started hearing voices and thought the CIA was putting something in his coffee to keep him awake. A few weeks later, on the way to the hospital to be treated, he threw himself out of a closed hotel window, ten floors up.

The cause of Olson's death came to light in 1973, along with the fact that throughout the 1950s and early 1960s, the U.S. government had been playing fast and loose with LSD. *Acid Dreams: The Complete Social History of LSD: The CIA, the Sixties, and Beyond,*[1] a 1985 book by Martin A. Lee and Bruce Shlain, tells how scientists, sponsored by the government, tested LSD on soldiers but also on civilians in psychiatric institutions and prisons. The subjects were given very little information, if any. Some weren't given the option to refuse. Some were given the drugs while tied to a bed. Some were dosed for six weeks at a time. Doctors even lobotomized patients, before or after taking LSD, in order to compare the effects.

The aim was to discover whether LSD could be used as a mind-control drug in espionage, as an instrument of psychological torture or as a weapon of warfare. And the underlying justification for these deeply unethical "studies" was the Cold War: the Soviets, it was said, would be prepared to do worse.

Similarly unethical experiments went on outside institutions as well. A CIA operative rented a flat in New York's Greenwich Village and installed two-way mirrors. He lured drunk, unwary people back and dosed them. There were so many bad reactions, he gave LSD a nickname: "Stormy."

When army interrogators used LSD on foreign nationals who were suspected of being spies or smugglers, one poor victim vomited three times. Another went into shock and then, when he was revived, said he wanted to die to escape the torture. When LSD was administered in intraspinal injections, it was described as producing an "immediate, massive, and almost shocklike picture with higher doses."

# THE UNDERPINNINGS OF PSYCHEDELIC THERAPY

Considering the way government researchers administered LSD, it's not surprising that an official view emerged that LSD made you mad. The researchers described tripping and psychosis as one and the same, and called LSD and similar substances "psychotomimetic"—that is, mimicking madness.

However, during the same era a new and opposite view of these drugs emerged. This came from psychiatrists and psychologists, who also wanted to find out how LSD could be used, but with the underlying aim of helping and treating people rather than manipulating or using them.

Dr. Humphry Osmond, a British psychiatrist based in Canada, was one of the pioneers of using psychedelics therapeutically. He gave LSD to patients to treat alcoholism. At first he was surprised when his patients described their experiences as insightful and rewarding, sometimes even as beautiful.[2]

Gradually, though, he became convinced that the name psychotomimetic was not the right way to describe these compounds. It was Osmond who, in 1957, in correspondence with Aldous Huxley, came up with the new name psychedelic, meaning "mind-manifesting."

In his early incarnation as a respectable Harvard psychologist, Dr. Timothy Leary was also a proponent of the therapeutic power of psychedelics. Like the government researchers, he conducted some of his research on prisoners, giving psilocybin to 32 inmates of Massachusetts Correctional Institute in Concord. The study aimed to find out if psilocybin therapy would help prevent recidivism (prisoners reoffending and being locked up again). But his study was very different from the government researchers': the subjects weren't forced or tricked into taking psilocybin. They were prepared in advance as to what the drug might feel like and allowed to provide input and criticism.[3] At the time, the study results showed that psychedelic treatment did reduce rates of recidivism although later analysis has disputed this.[4]

The two opposing views of psychedelics—they make you psychotic versus they are therapeutic—persisted throughout the 1950s and 1960s. But despite growing evidence from hundreds of experiments on the therapeutic side, the psychosis claim won out in the end. The media and government in both the U.S. and the UK followed the line that psychedelics were too dangerous to be left uncontrolled. This then justified them being banned at the end of the 1960s.

But the therapeutic studies were not wasted. This first wave of research, the hundreds of studies conducted during the 1950s and 1960s, now underpins modern research including the work we do at Imperial College. Although these studies were forgotten, belittled or ignored for many years, a lot of the principles and methods developed by the early researchers, as well as their findings, still inform the way we do psychedelic therapy in trials today. And their data, showing that the drugs are very safe when used under medical supervision, is very reassuring.[5]

## WHAT MAKES PSYCHEDELICS A FORCE FOR GOOD?

How could some subjects appear to go psychotic after taking psychedelics, and others report great benefits?

Professor Stanislav Grof, one of the first people to develop psychedelic therapy, called psychedelics "non-specific amplifiers of the psyche." He meant that whatever someone is feeling before a psychedelic experience will likely intensify once they've taken the drug. And so the subjects who were prepared for the drug and supported throughout by a therapist had a good experience, but those who weren't prewarned or didn't know what was happening to them did not. The army and CIA studies often took place in a sterile lab or on subjects who were tied to a hospital bed or already drunk, and they gave subjects zero support.

We now know that the preparation of the person, their mindset and the physical environment are key to the result. All this makes up the "set" and the "setting," mentioned earlier, labels that originated in this first wave of research. In fact Leary wrote a series of papers on why getting this right is crucial, and how to do so.[6]

Leary defined the set as the subject's mindset and all the internal factors that affect the drug experience. This includes personality and how suggestible the person is, how much they can "let go" after taking the drug. But it also includes the person's assumptions about the drug's effects, as well as their "mood, expectations, fears, wishes."[7]

The setting is the environment, which includes the room or place you're taking the drug, the music and lighting and time of day. It also includes the social and cultural environment, i.e., whether the drug is legal, acceptable or stigmatized, seen as recreational or healing.

An illustration of this is that nearly half of people who took psychedelics in the 1960s reported having had a bad trip.[8]

Martin A. Lee and Bruce Shlain in *Acid Dreams* state: "The high percentage was in part a consequence of the widespread anxiety that ensued after LSD was declared illegal in late 1966."[9] The laws "created a hostile environment that predisposed people toward more traumatic reactions," aka a bad cultural setting. That is why, they say, as the political anti-LSD climate became less intense in the 1970s, the percentage of bad trips went down.

When LSD was legal, early studies stressed the importance of the right surroundings. The British psychiatrist Dr. Ronald Sandison created the prototype for modern psychedelic-assisted therapy. He used a non-psychedelic dose of LSD to help patients who'd become "stuck" in psychotherapy, a technique called psycholytic psychotherapy. In 1955 he built the world's first purpose-built room for LSD therapy at Powick Hospital in Worcestershire, equipped with a couch, a chair and a blackboard for the subject to draw on. However, Sandison's strong dislike of the recreational use of drugs and the media attention that came with it led him to stop working with LSD in 1964.[10]

In studies at Imperial College, we are careful to prepare people thoroughly on what to expect, we make the room welcoming and the lighting relaxing, and we have a customized playlist. We also make sure they have two therapists or guides with them throughout the trip, which is currently the standard protocol in most psychedelic trials. When MDMA-assisted therapy for PTSD is approved, it will also include two therapists, although this may not end up being the case for all psychedelic therapies.

## WHAT DOES A PSYCHEDELIC THERAPIST DO?

Grof wrote about psychedelics being tools that, used in the right way, enable an outcome rather than cause it. "Is a knife a terribly dangerous thing or is it a useful thing?," he asks. While a police chief might describe a knife used by a criminal to commit crime, a chef will describe a knife used to make delicious food. The idea is that psychedelics aren't inherently dangerous when used in the right way.

Sandison used LSD in therapy as a tool to help break down his patients' resistance to talking about problems, help them retrieve key memories and make them more responsive to therapy. Alongside the perceptual, visual and auditory changes, mystical experiences and depersonalization, the qualities that make psychedelics good therapy tools are that they increase empathy, openness, trust and suggestibility.[11, 12]

Ketamine and MDMA have these qualities too; they open the person up, allow them to go into their emotions. These drugs make instant changes to how the person thinks and feels that, with the help of a skilled therapist, can become long-lasting.

Some experts theorize that the drug is the treatment—a kind of trip switch for the way the brain works—and the therapist is there to make sure it's safe. Others say that the therapy is the treatment, and the drug helps open the person to their emotions, go to places in their psyche that they wouldn't normally go.

It's more likely that the unique properties of these drugs—producing changes in established ways of thinking plus neuroplasticity—combine with the guidance of good therapy to produce insights and long-lasting changes. Rather than acting like a trip switch, it's probably more accurate to say that the brain is being rewired.

In the U.S., ketamine is now prescribed both with and without therapy. There are no head-to-head studies comparing the use of ketamine in depression with and without therapy. But the University of Exeter KARE study with alcohol-dependent people (see the next chapter) found ketamine plus psychotherapy was better than ketamine without therapy.

At Awakn clinics, where we do ketamine-assisted therapy, we have found that patients do want to talk about these powerful experiences. The therapist's job is to help the subject get the most healing out of the experience while reducing the possibility of harm. And a therapist can guide the patient through any difficulties, such as paranoia, fear or anxiety. This is especially important if the subject is vulnerable, for example, has a mental health condition such as depression or dependence.[13]

Another pioneer, psychologist Dr. Bill Richards of Johns Hopkins medical school, author of the book *Sacred Knowledge: Psychedelics and Religious Experiences*, has worked in psychedelic therapy since 1963. He says the role of the therapist in psychedelic therapy is unique, completely different from the usual role of a therapist in psychotherapy. The psychedelic therapist acts as more of a guide, "providing presence, patience, receptivity in a safe container—a safe therapeutic relationship where there is genuine trust, genuine caring, genuine confidentiality and safety."[14] And he says the therapist isn't there to stop the subject going into difficult emotional places, because that is part of the healing experience, but they will be by their side while they are there. Both during the trip and afterward in the integration sessions, they can help answer questions about what is coming up for the subject and help them make sense of new images, memories, emotions, physical sensations and thoughts. And they can help the subject move in and through all these.[15]

Most psychedelic-assisted therapy is about letting the patient follow their own insights and self-discovery rather than the therapist intervening. Taking a psychedelic with this kind of minimal guidance, Richards says, "we witness...a certain wisdom of the patient's mind, manifesting, choreographing content in ways that are perfectly designed for that particular person at that particular time." In other words, it is the patient or subject who leads the recovery. The therapist doesn't give advice or teach, rather asks questions that help the subject to find their own ways through their difficulties.

## THE STAGES OF PSYCHEDELIC THERAPY

Clinical psychologist Dr. Rosalind Watts, formerly at Imperial College, now runs ACER Integration (Accept, Connect, Embody, Restore), an online community for people who have taken part in psychedelic therapy.[16]

She was the lead therapist for the Imperial College trial that compared psilocybin and escitalopram. When I spoke to her on the *DrugScience* podcast, she described what happened during the days of the trial.[17]

### Day 1: Prep Day

The purpose of the prep day was to maximize a sense of trust and rapport, for the participant to get to know the guides in an authentic, human way. Being in the altered state of the psychedelic experience is very intense and it can be quite frightening. The subject needs to know their guides are human, to trust that they will be ethical while the subject is vulnerable and help them when needed. One of the ways we did this was to have lunch together.

We would discuss what we'd do in different situations that might come up, for example, if they wanted to give permission to the guides to hold their hands. Participants would often have lots of questions about their fears, such as "What if I embarrass myself?" or "What if I lose control?" We'd work through these, and how we'd respond as a team to each of them.

The afternoon session was a visualization, a practice run for the next day. We'd ask the subject to imagine the psilocybin experience as like going on a pearl dive. To visualize swimming out to sea and diving down into the water. Rather than getting distracted by all the pretty colors, we'd ask them to dive down to the darkest layers, to find the deepest part of themselves. We'd also play the soothing, relaxing music that would be used the next day.

## Day 2: Trip Day

We presented the capsule to the subject in an engraved wooden bowl, a little ceremony. Once they'd taken it, the subject would lie back on the bed and put on their eyeshades and headphones. The playlist played both through the headphones and into the room. The two guides would sit either side of the participant and often, the participant would reach out and hold hands with both the therapists.

Most of the session day was quiet. Sometimes, the subject would want to engage with the guides, and we'd be with them for whatever came up. We'd make a judgment whether it was an important conversation, in which case we'd keep talking to them, or if they were avoiding something difficult, when we'd encourage them to put their eyeshades back on and go back in.

Most trips contain both good and bad. Sometimes, people would get stuck in a painful or dark place, which might have been terrifying alone. But with two guides, it is a rich opportunity for the subject to access the darkest corners of themselves, and to look at these together.

At the end of the five- to six-hour session there was a little time to talk. Then the subject would go to the accommodation next door to sleep.

## Day 3: Integration Day

At 9 a.m., I'd go out to the reception area to get the subject. When I saw them, I'd always know how it had gone. Sometimes their face would look relaxed, glowing, even ten years younger. It was amazing to see the change in them from one day to the next. Sometimes they would be disappointed, because they wouldn't feel any different.

We'd start with a special cup of delicious caffeine-free chai. Integration is a relaxed, fluid day, where the participant can talk through anything that has come up. Often people would have some confusion to talk through, some things to clarify. Sometimes they'd need more grounding, then we might listen to the playlist to help them with this, or do some meditation. They would talk through pearls they might have gathered, and how they might continue to take these into their behavior, thinking and life.

## SAMPLE SCHEDULES FOR DIFFERENT PSYCHEDELICS

There are differences in the way therapy is done depending on the substance and what's being treated. These are some examples; different organizations and studies will have protocols that diverge from these guidelines, but they often go broadly along these lines.

* **Psychedelic therapy**
  Day one: preparation. Day two: drug treatment. Day three (or later) integration sessions.
* **MDMA therapy**
  One to three therapy sessions, and a preparation session specifically about MDMA. One to three drug treatment sessions, followed by integration sessions as part of a longer psychological treatment course.
* **Ketamine therapy**
  Eleven sessions in total, including four drug treatment sessions. More information at awaknclinics.com.
* **Ayahuasca therapy (at a retreat)**
  In an eight-day retreat, there may be two to four ayahuasca ceremonies.
* **Ibogaine therapy (at a retreat)**
  In an eight-day retreat, there may be two iboga ceremonies.

## MUSIC: THE THIRD THERAPIST

Music plays a central role in psychedelic therapy because, when a person's usual control over the senses breaks down, it leaves a freer, less inhibited way of processing sounds. Music can change someone's whole experience, enhancing the vividness of their mental imagery, activating emotions, thoughts and memories. People feel music more intensely under psychedelics, but it can also take them on a journey, a living dream of personal content, which can enhance the effectiveness of the therapy.[18, 19]

The playlist for the first Imperial brain imaging trial was designed by a former PhD student of mine at Imperial College, Dr. Mendel Kaelen, now founder and CEO of Wavepaths, a company that researches and develops the psychotherapeutic potential of music.

He says: "It's not an exaggeration to say that one song can be the source of a therapeutic breakthrough, a life-changing set of emotions and insights."

Before the study, Kaelen looked at existing playlists for trip therapy, which tended to be classical music, such as Beethoven's Piano Concerto Number 5. The playlist he developed was made up of more current music and aimed to minimize any religious associations while supporting the peak experience. It included ambient, contemporary classical and traditional/ethnic music styles. It supported each phase of the psychedelic experience, as defined by Grof.[20]

The phases are, in order: pre-onset, onset and building toward peak, peak, reentry and return. Most of the playlist is calm, except for at the peak, when the music alternates between evoking strong emotions and being soothing and reassuring.

A week after the first depression study, the subjects were questioned on how the music affected them.[21]

People said that it intensified their experience, including enhancing and intensifying their emotions. For example, one subject said: "Normally when I hear a piece of sad music or happy music I respond through choice...but under psilocybin I felt almost that I had no choice but to go with the music. [...] I did feel I was being held. And it did feel like the music opened [me] up to grief, and I just was very happy for that to happen."

Sometimes the music was welcome; sometimes it intensified emotions they did not want to feel, such as "fearfulness," "sadness" or "fear." People also described music as a source of guidance, support and calm, discomfort and irritation and mismatched with their inner experience.

## THE FUTURE OF THERAPY? MEDITATION, MINDFULNESS AND PSYCHEDELICS

When we published our first paper showing that psilocybin switched off the ACC in the brain, psychiatrist Dr. Judson Brewer emailed me to say, "Why didn't you quote our work on meditation switching off the DMN?" Brewer is based at the Mindfulness Center at Brown University in the U.S., where they have looked at the brain basis of meditation. The study he was referring to had come out a few months before ours. It imaged the brains of experienced meditators and beginners across several types of meditation. And it showed that meditation switches off parts of the DMN in a similar way to psilocybin.[22]

The truth was that our team hadn't known brain imaging

work was being done on meditation, but we, and many others, quickly realized that this convergence of results was important. Soon there was a joke going around: You can achieve nirvana through 20 years of meditation training... Or you can achieve it through 20 minutes of psilocybin!

Johns Hopkins have started to work in this area, including a study that used psilocybin to accelerate the acquisition of meditative skills. The idea is that, by putting the brain into this state with psychedelics, you can help people acquire the necessary skills to get into and persist with deep meditation.[23]

One study compared a group that was given two very low doses (1mg or placebo) of psilocybin plus meditation/spiritual training to groups given high doses plus training. Six months later, the people in the high-dose group were more likely to have continued their spiritual practice, and to show lasting positive effects from the psilocybin. Professor Franz Vollenweider's research group in Zurich has had similar results.[24]

Some researchers are looking at mindfulness to promote change during the post-psilocybin period.[25] Mindfulness is the latest approach to treating mental health conditions. Previously, the standard UK NHS approach was cognitive behavioral therapy (CBT). In simplistic terms, this therapy is based on the premise that you can use your rational mind to change your thinking. But an approach codeveloped by Professor Mark Williams, mindfulness-based cognitive therapy (MBCT), has moved this on, showing that the answer does not lie in reasoning yourself out of your thoughts but in detaching yourself from them.

It may be that the synergies between mindfulness and psychedelics may help mindfulness work better and so may give us a major new way to treat depression, addiction and other mental health disorders. A review looked at the effects of both mindfulness meditation and psilocybin and concluded that they do have some similarities in effect—such as on mood and neuroplasticity—as well as differences. But it also found that the two may complement each other, i.e., meditation may help bring on or prolong the psilocybin peak experience.[26]

## THE ETHICS OF SET AND SETTING

Psychedelic therapy takes elements from what was tried and tested by researchers in the 1950s and 1960s, as well as from non-psychedelic therapeutic approaches and from indigenous ways of using plant medicines.[27]

It may be that we could learn better ways to use psyche-delics therapeutically from past and indigenous cultures. Because it's likely that cultures who've used them for hundreds or thousands of years have developed best practice, according to anthropologist Dr. Michael Winkelman at Arizona State University.[28] We have adopted the shaman or guide, and the importance of music, but psychedelic healing practices also often include fasting, drumming, dancing, chanting and/or singing, as well as being communal. It seems likely that the collective nature of, for example, an ayahuasca ceremony or the ayahuasca churches adds to its power.

The fact that our modern use of psychedelics owes a lot to indigenous cultures brings up a whole range of ethical questions around origins.

While the expansion of the global psychedelic marketplace

makes it now inevitable that people will access plant medicines outside their original countries and contexts, the companies that are profiting from the plants and the indigenous knowledge of them should give back to the originators, as well as protect the land they come from.[29]

For example, the boom in ayahuasca tourism in South America has led to reports of ayahuasca plant sources becoming scarce, and plants being stolen from the jungle. The peyote cactus is now endangered, as is the *Incilius alvarius* toad, the source of 5-MeO.

Now, organizations such as the International Center for Ethnobotanical Education, Research, and Service (ICEERS) are helping protect the cultural capital and land of indigenous people in areas where ayahuasca and ibogaine originate. The focus on protecting land is especially apt because psychedelic experience has been shown to increase the feeling of being connected to nature.[30]

Some people argue that plant medicines are only or most effective when they are grown in the traditional place, given by indigenous shamans in the traditional way and in the place where they came from. There are now discussions about making ayahuasca an international trademark, in the same way that champagne can only be made in the Champagne region of France.[31]

## WHAT HAPPENS IF THERAPY GOES WRONG?

There are rules and boundaries that all therapists and mental health professionals should keep to in terms of their relationship with a patient or trial subject. On top of this, it's recognized that someone who's taken a mind-altering drug is in a unique state of vulnerability, being more open and/or suggestible.[32]

During a MAPS MDMA study in 2014, two therapists acted against MAPS guidelines. They were a husband-and-wife therapy team, working with a patient who had PTSD because of sexual assault.[33] Video of one session shows the two therapists pinning the patient to the bed, cuddling and blindfolding her.[34] At the end of the study, the male therapist continued working with the patient, and later started a relationship with her, including having sex. In 2018, the patient made an allegation of sexual assault.

The fact that the therapists acted unethically and unprofessionally doesn't make the study results invalid. The fact that it happened, though, does highlight a possible risk with all kinds of psychotherapy: that therapists do sometimes have sex with clients. A recent survey found that seven out of ten therapists have found a client sexually attractive and 3 percent have had sexual relations with their current or former clients.[35]

It is important that these transgressions came to light. Both during trials and treatments, psychedelic therapists must be held to the highest ethical standards. At Imperial College we film the therapeutic sessions as part of our safeguarding, in case any patient does make a complaint, or if a patient just needs reassurance that professional boundaries were kept.

## DO PSYCHEDELICS HEAL WITHOUT THERAPY?

The short answer is: no, not effectively or safely. That said, evidence from surveys and case studies suggests that taking psychedelics for nonmedical reasons can lead to life changes.

One survey by Imperial College suggested that even two years after taking a psychedelic, people were still enjoying improved well-being.[36] Another survey showed that taking psychedelics outside therapy has an antidepressant effect that lasts at least a few weeks.[37]

In a third survey of over 2,500 people, the majority reported reductions in anxiety and depression, and improvements in emotional well-being. It's interesting that those who'd taken ayahuasca, which usually involves ceremony and talking as well as taking the drug, reported slightly higher levels of well-being. However, 13 percent also reported at least one harm, believing their psychedelic use had led to more smoking, use of other drugs and, for a very few, suicidal ideation.[38]

It's never advisable to take a psychedelic without support, and especially if it's to help any mental health issues. In studies at Imperial College, we've seen that people with depression often have a harrowing, difficult time during trips. That doesn't stop them feeling that it was worth doing, but they really do need the support of the two therapists. Also, not everyone is suitable for psychedelic-assisted therapy (see Who Is Excluded from Psychedelic Trials? on page 288).

## SOME PSYCHEDELIC EXPERIENCES THAT CAN BE DIFFICULT

This list of examples comes from MAPS.[39]

- The feeling/experience that one is going crazy or losing one's mind, or that this will never end.
- Energetic experiences. People go through powerful releases,

rendering their bodies out of control, shaking, twisting and vibrating.

- Old traumas can be remembered and relived. For example, physical trauma such as reliving one's birth; childhood abuse and/or illness; memories of famine and/or war; accidents; rape. Or intellectual, emotional trauma, such as reliving verbal abuse, a lack of basic emotions or body contact, love or nurture, or a disassociation due to a traumatic experience.
- Other possible experiences include:
  - Remembering different deaths.
  - Reliving drowning, torture and many other physical experiences from this and other lives.
  - Reliving mystical states.
  - Identifying with and reliving in detail the victimization of humans throughout history.
  - Leaving the body and traveling in the spirit realm.
  - Merging with rocks, animals and plants and experiencing the pollution and death of the planet and different species.
  - Merging with people, reading their minds, feeling their emotions.
  - Being caught in a certain experience.
  - Having a UFO experience.
  - Being overwhelmed by feelings and emotions.

## WHAT IF I HAVE A BAD TRIP?

Even a challenging trip can result in improvements in mental state. People who have taken part in ayahuasca ceremonies or used ibogaine for addiction (see Chapter 7) often say the actual trip can be very frightening and unpleasant. However, they also often report an enduring sense of well-being and inner harmony that lasts for months or even years. This can make the transient discomfort of the actual bad trip worthwhile. It's very likely (though not easy to study) that the same switching off of the DMN occurs whether you have a good or bad trip. The possibility of a bad trip does underline that you need to do this treatment in a licensed clinic or in a trial.

Psychedelics must not be treated lightly. Therapeutic trips can be very challenging, leading to insights and the recall of memories that can be disturbing as well as insightful. For these reasons self-medication, and especially going it alone without a supporter who is not tripping, should be avoided. But because psychedelics are so powerful, they can be harnessed for good too. In fact, they need to be powerful in order to help people recover from resistant depression with its deeply entrenched negative thought loops.

# Chapter 7

# CAN PSYCHEDELICS TREAT DEPENDENCE?

Suddenly the room lit up with a great white light. I was caught up in an ecstasy which there are no words to describe. It seemed to me in my mind's eye that I was on a mountain and that a wind not of air but of spirit was blowing. And then it burst upon me that I was a free man... now for a time I was in another world, a new world of consciousness... and I thought to myself, "So this is the God of the preachers!" A great peace stole over me.[1]

**THIS IS A** not untypical description of a psychedelic trip, you might think. What makes it remarkable is the author: Bill Wilson, the founder of Alcoholics Anonymous. An alcoholic at the time, Bill had this breakthrough while undergoing the Belladonna Treatment, which was popular in the 1930s. It involved sudden withdrawal from alcohol plus a combination of drugs similar to scopolamine (which we met in Chapter 3) including one called "insane root," as well as hypnotics including chloral hydrate and morphine, plus some strychnine. The cure's other name was "puke and purge."[2]

Today, we don't use belladonna-derived drugs; they can lead to seizures and cardiac problems. But in the 1930s this cure was a step up from the usual cold-turkey approach.

Wilson's psychedelic experience changed his life. He never drank again. And he went on, of course, to affect the lives of the millions of people who've stopped drinking with the support of AA.

Bill's life before his transformation was typical of the many millions of young men who become alcohol dependent. He came from a family with a tendency to alcoholism—his paternal grandfather was an alcoholic—and as a child, he suffered psychological damage when both his parents abandoned him. Although Bill was talented and intelligent, he was also socially very anxious and vulnerable to bouts of depression.

At Yale University, Bill started to drink to deal with his anxiety attacks, finding that alcohol liberated him from his shyness. After his first experience with alcoholic cocktails he wrote, "I had found the elixir of life," despite having got severely drunk and blacked out. Of course, this was not a particularly untypical event as he pointed out: "But as everyone drank hard, not too much was made of that." Over the next decade, though, his binges became more and more problematic. He missed graduating as a lawyer because he was too drunk to turn up to receive his diploma.[3]

Bill messed up several jobs due to drunkenness and bingeing. He was committed for treatment with a New York expert, Dr. William Silkworth, at the Charles B. Towns Hospital, an expensive uptown drying-out facility. Dr. Silkworth viewed alcoholism in a similar way to how psychiatrists like me view it today: as an obsessional desire or compulsion to drink plus the inability to stop drinking once the person has started.

Like many people with alcohol problems, Bill found it encouraging for his alcoholism to be treated as a medical condition rather than a moral weakness. But he still couldn't stop. He realized that if he didn't, he would either die from his alcoholism or, the usual practice of the time, be committed to an institute for inebriates.

It was during Bill's fourth detox that he underwent the Belladonna Treatment and stopped drinking. Along with his old drinking buddy Ebby Thacher he began to develop the 12 Steps and other concepts that still underpin AA. For example, he referred to Ebby, who'd encouraged him into treatment, as his "sponsor."

Ebby had recovered by finding God and belonged to an evangelical Christian group. He told Bill that faith would liberate him too. Lying in his hospital bed, deep in the misery of alcohol withdrawal, Bill said to Ebby, "I'll do anything! Anything at all! If there be a God, let Him show Himself." It was at this moment that he had his vision of God.[4]

When people are suffering from delirium tremens (the DTs), they can have vivid and horrifying visual hallucinations, often of spiders and insects. However, most alcoholics are not cured by their withdrawal reactions, even if they do hallucinate like this. It seems clear that Bill's change in behavior was down to his experience on consciousness-changing drugs, not just his alcohol withdrawal.

Now, in the new wave of research, the use of drugs that change consciousness to treat dependence is back. After depression, the next most researched and likely future use for psychedelics is in treating a whole range of substance addictions: alcohol but also tobacco, heroin and cocaine.

This is such important work. Addiction is a major health problem. In the UK alone, the total cost of drug misuse to society is over £19 billion[5] and the cost of alcohol misuse is estimated to be between £21 billion and £52 billion.[6] This is more than the cost of treating cancer (£18 billion)[7] and cardiovascular (£16 billion) disorders put together.[8]

It's extremely hard to stay abstinent from an addiction such as alcohol or drugs. Even if someone is treated in a residential detox center, which will give them the best chance of success, when they come back to the "real world" and see their old mates they're very likely to fall back into their old way of life.

There aren't currently many medicines available to help people. But if psychedelics continue to perform as well in the real world as they do in studies, they may change and even save millions of lives. Not only of the addicts themselves, but of their families and children too.

## A BIGGER PICTURE

Bill Wilson's experience brings up a fundamental question about psychedelic therapy and addiction. It's one we still don't know the answer to: Is it just the trip itself that changes you? Or is it that the trip allows you to engage in behavior that protects you against relapse?

Perhaps the biochemical changes in his brain due to the psychedelic ingredients in the Belladonna Treatment enabled Bill to make the profound and enduring changes in his attitude and being that allowed him to become abstinent. Like many people who have a psychedelic experience, Bill gained a sense of the existence of something more significant and bigger than alcohol or himself. For him, this was God.

Or perhaps the changes he was able to make after taking a psychedelic—joining an evangelical Christian group and finding solace in God and the group, finding the purpose and strength to start AA—allowed him to stay abstinent.

Twenty years later, Bill's AA story got even more interesting. In the 1950s, when LSD was legal, Bill and his wife took part in an experimental group LSD session, along with Aldous Huxley, the psychotherapist and LSD pioneer Betty Eisner and Bill's spiritual adviser, the Jesuit priest Father Ed Dowling.

On LSD, Bill relived his first psychedelic experience. He said it helped him overcome the barriers of the self, or ego, that stopped him from having a direct experience of the cosmos and of God.[9]

Bill came to believe that using LSD, in a carefully structured way and a controlled setting, could become part of recovery for all alcoholics. He said, "I consider LSD to be of some value to some people, and practically no damage to anyone." And his influence contributed to the U.S. government funding six clinical trials to test LSD as a treatment for alcoholism.[10]

When LSD was made illegal in the U.S. in 1967, Bill continued to advocate for its use as a medicine. But the AA stayed antidrug. It still is, including drugs that are therapeutic agents for the brain, such as lithium. Even in the present day AA preaches that a "spiritual awakening" is key to recovery while rejecting how its founder achieved his.

The LSD ban was opposed by some leading and open-minded politicians such as Bobby Kennedy, whose wife Ethel reportedly had LSD therapy at Hollywood Hospital in Vancouver. He said: "If [clinical LSD projects] were worthwhile six months ago, why aren't they worthwhile now? We keep going around and around... If I could get a flat answer about that I would be happy. Is there a misunderstanding about my question? I think perhaps we have lost sight of the fact that LSD can be very, very helpful in our society if used properly."[11]

## THE HIDDEN HISTORY OF LSD

In 2012, the six historical trials on LSD and alcoholism were re-analyzed using a modern statistical approach, meta-analysis.[12] The analysis found that LSD was at least twice, if not three times, as effective at reducing relapse rates than any other proven and licensed treatment for alcoholism developed since (although there are very few). And this is even more impressive because it uses a very high bar for success: being abstinent.

At the time I was the editor of the journal this paper was published in, and it was the first time I'd ever heard of psychedelics being used to treat addiction. I hadn't heard of them at Oxford when I was training in psychiatry, or at Guy's Hospital in London when I was learning about addiction as a medical student, or even when I was working at the National Institute on Alcohol Abuse and Alcoholism in the U.S.

This is because in order to make sure LSD would be—and stay—banned, politicians knew it had to be discredited. Not only did it become much harder for researchers to obtain, it became unacceptable for doctors or researchers like me even to consider studying it. Politicians didn't want the truth to be known because it would challenge the ban. And so those LSD studies, and others like them, were written out of history.

Using a back-of-an-envelope calculation, since LSD was banned in the late 1960s possibly 100 million people have died prematurely from alcohol abuse. They were denied a treatment with some strong evidence. And even if LSD had helped only 10 percent of them, that would have saved 10 million lives.

The meta-analysis ignited my interest in the potential of psychedelics. I read about Bill Wilson and his involvement in the alcohol studies. I began to learn about other forgotten studies too. One 1973 study that used LSD to treat heroin addicts, a notoriously hard-to-treat group, had particularly impressive results.[13]

The study took 78 volunteer heroin-addict inmates from Maryland prisons. The group was randomly split: Half took one high dose of LSD during six weeks staying in a halfway house. The rest attended an outpatient clinic including daily urine monitoring and weekly group psychotherapy. Apart from the halfway house, both groups were treated the same.

After a year, almost none of the control group had managed to stay off heroin. But around 25 percent of the treatment group had. This is extremely impressive. To put it into perspective, at both the Bristol and Imperial labs we have had funding to look at brain change in heroin addicts after three months of abstinence. We couldn't find a single subject who'd managed the three months.

# THE BEGINNINGS OF MODERN ADDICTION RESEARCH

The first addiction study of the new wave of psychedelic research was also the first ever study to use psychedelics to treat tobacco addiction.

This is likely because in the 1950s and 1960s, this addiction wasn't known as the killer that it is. Currently, tobacco kills more than 8 million people a year worldwide.[14]

The study's author is Matthew Johnson, professor of psychedelics and consciousness at Johns Hopkins University. He chose smoking because he saw that psychedelics had a "cross-substance" effect; that is, they work for multiple addictions. Also, there are definitive blood and breath tests that prove whether someone's been smoking or not, an advantage when it comes to addiction as people can sometimes be less than honest.

Johnson recruited 15 hard-core smokers, who'd smoked an average of a pack a day for 30 years. Over four months, the smokers were given two to three psilocybin trip sessions plus sessions of cognitive behavioral therapy (CBT), the gold standard therapy for quitting.

The results were remarkable: One week post the first session, nobody was smoking. After six months, 12 people (80 percent) were not smoking. For the best alternative treatment—a drug called Champix (varenicline)—this percentage would be only 30 percent.[15]

And the results stayed good: After 12 months, 10 people (67 percent) were still not smoking. And at a long-term follow-up (16 to 57 months), nine people were still abstinent.[16]

Johnson is following this up with a larger study, still under way at time of writing, comparing one session of psilocybin with nicotine replacement therapy.[17] And in 2021, he received the first federal grant to research psychedelics in over 50 years, from the National Institutes of Health's National Institute on Drug Abuse.

Johnson's research shows that psychedelics have the potential to revolutionize quitting smoking. If psychedelics hadn't been banned, it seems likely that this use would have been discovered much sooner, and many millions of lives would have been saved.

## CAN YOU USE A DRUG TO TREAT A DRUG?

Experts assume that because these drugs are controlled, they are dangerous and so shouldn't be used as medicines.

But in fact, we already use "dangerous" drugs as medicines, such as heroin and other opiates. And we treat addictions to those substances with medicines such as methadone or buprenorphine, which although they aren't as dangerous as the illegal addiction do still carry a high risk of addiction.

However, the main point is that using psychedelics is not replacing one addiction with another, as psychedelics are not addictive (see Chapter 13).

MDMA does have potential for addiction, but it is low, and even lower when used in a medical setting. Ketamine has a higher risk of addiction, but the fact that it's used just a few times during a treatment program makes the risk of dependence also very low (for more on risks, see Chapter 13).

## WHY DO PSYCHEDELICS WORK FOR ADDICTION?

Addiction begins as a behavior that's enjoyable or provides relief, such as cigarettes for stress or alcohol for social anxiety or cannabis for insomnia. But eventually it becomes a self-sustaining habit. The positive effects of addictions are laid down as deep-seated memories that link the location, people involved and experience of the addiction with the positive emotional effects.

Once addiction sets in, choosing to stop is very difficult, often impossible. The urge to engage in the behavior becomes so powerful that it interferes with normal life, often to the point of overtaking work, personal relationships and family activities. This quote from a patient of the physician and addiction expert Dr. Gabor Maté sums it up: "When I was using, I had tunnel vision."[18]

Anyone who's met an addict knows that most addicts don't want to take drugs or drink but are compelled to by something that's beyond them.

## THE WHOLE-BRAIN EFFECT

Most existing medications that we use in addiction work at the level of a specific neurotransmitter. For example, the drug naltrexone blocks heroin from binding to opiate receptors, taking away the drug's effect. It treats alcohol dependence by blocking some of its effects too, so reduces bingeing. However, even though this prevents some of the addictive effects, it won't "cure" addiction or even deal with all the symptoms.

Psychedelics have a broader mode of action. Because of this, they are the only drugs we know about that have the power to fundamentally change what, for many people, is a decades-old behavior and associated lifestyle. Potentially, they can cure people.

They do this by disrupting the brain processes that underlie addiction. There are at least four kinds of these and different people with different addictions will have different degrees of each one. The four are: stress sensitivity (you can't bear stress), motivation to use (you want it), impulsivity (you start doing things too fast) and compulsivity (you can't stop).

The diagram in Figure 7 of the nonaddicted brain shows the current thinking about how people control their urges to do anything, but especially to resist drugs and other addictions.

There are four separate brain circuits, which work together to regulate behavior.

1 DRIVE/MOTIVE
   Nucleus accumbens/ventral pallidum. This is the motiva-tion and drive system. It works on rewards and predictions of pleasure. In addiction, this system can become compul-sive or habitual.
2 EMOTIONS
   The hippocampus/amygdala. This is the center of learning, emotions and the memory of emotions. If you have high sensitivity to stress, you can find emotions hard to bear and so use substances to suppress them. In the end, addictive behaviors can themselves become very stressful.

3 DECISION-MAKER

Prefrontal cortex (PFC). This is your cognitive control, the top decision-maker of your brain. This is where you make your decisions about what you should or shouldn't do. It's also where you control your impulses.

4 ACTION TAKER

Orbitofrontal cortex (OFC). This is the final decision point of actions. It's controlled by the PFC. If your PFC decides you're going to have a drink of water, the OFC will do it. If your PFC says not to do it, your OFC will not act.

The diagram in Figure 7 shows the feedback between these areas of the brain. The size of the arrows indicates how much influence they have over one another.

In the balanced, nonaddicted brain, the PFC controls the final decision-making.[19]

Figure 7: Schematic of the balance between the four key brain regions involved in addiction. In the nonaddicted state the PFC regulates decision-making.

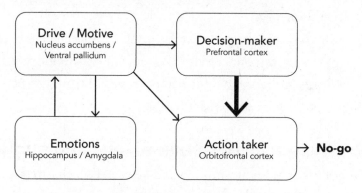

Source: Adapted from Baler R, Volkow N, 2006, *Trends Mol Medicine* 12(12): 559–566. doi: 10.1016/j.molmed.2006.10.005.Epub Oct. 27, 2006.

In the addicted brain, things work differently. How this happens—how you end up doing something you don't want to do—can depend on the addiction.

In certain addictions, for example, cocaine and, to some extent, alcohol, the drive/motive system is enhanced, leading to the OFC escaping control of the PFC.

In some addicts, we think the PFC loses its capacity to control decisions, possibly as a result of drug use. And so, the OFC is less constrained.

In someone with PTSD, the emotional drive (the connectivity between the emotional and drive/motive regions) is overactive and so they may drink to suppress it.[20]

In recent research from my group at Imperial, we scanned the brains of alcoholics and nonalcoholics to look at the connectivity between the emotional and drive/motive regions.[21]

It showed that alcoholics do have more connectivity in this circuit than nonalcoholics. This gave the first proof in humans of the theory that enhanced emotional drives are associated with alcohol misuse.

The research also showed that the longer someone had been drinking, the more connectivity they had, i.e., the more well ingrained this addiction circuit had become. This may explain why long addictions are harder to quit.

## HOW DO PSYCHEDELICS WORK AGAINST ADDICTION?

The theory is that ketamine (covered later in this chapter) and the classic psychedelics disrupt the overactive links between drive/motive and the OFC. This allows the PFC to regain control.

## Figure 8: In the addicted state, the balance between the four key brain regions is disrupted.

The bold arrows represent more activity. In the addictive state the drive/motives and the emotional systems override the decision-maker, which overstimulates the action taker.

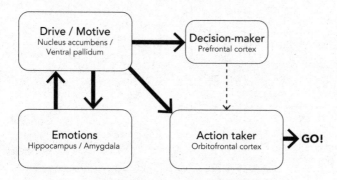

Source: Adapted from Baler R, Volkow N, 2006, *Trends Mol Medicine* 12(12): 559–566. doi: 10.1016/j.molmed.2006.10.005.Epub Oct. 27, 2006.

As you read in Chapter 5, on depression, imaging studies show the brain's activity usually has a regimented, synchronized rhythm. But psychedelics fundamentally disrupt that. For a period, they free the brain from its old ways of thinking, such as repetitively ruminating about an addiction. This allows new ways of thinking to take hold. MDMA (also covered later in this chapter), appears to work in a different way by suppressing the excessive emotional drive, reducing the connectivity between the emotional and drive/motive regions. By stopping and/or reducing fear and/or anxiety responses, the brain balance is restored.

On top of these effects, all these drugs facilitate neuroplasticity, so can lead to enduring changes in brain connectivity. This means that, when people get new insights and new ways of thinking, including the strategies they learn in therapy, these are laid down better and so can be used in the future.

## Figure 9: How psychedelic drugs and MDMA restore balance

Psychedelics and MDMA interrupt the overactive systems, restoring the central role of the decision-maker over use.

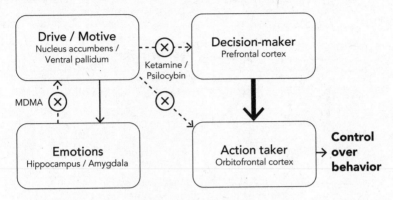

Source: Adapted from Baler R, Volkow N, 2006, *Trends Mol Medicine* 12(12): 559–566. doi: 10.1016/j.molmed.2006.10.005.Epub Oct. 27, 2006.

# NEW WAYS TO TREAT ALCOHOL DEPENDENCE

In the 1980s the Soviet Union had (and in fact, Russia still has) the highest consumption of alcohol in the world. So it was a fitting place for the rebirth of psychedelic addiction treatment.

Professor Evgeny Krupitsky is a true pioneer. He began to research ketamine because it was both available and legal, and he thought it might be an interesting way to disrupt the psychological processes of addiction. His results were impressive: after a year, half the people in his study had stayed entirely abstinent, after just a single session of ketamine.[22]

However, Krupitsky's research didn't get the recognition it deserved from addiction experts either in Russia or around the world. This was probably down to timing. By the late 1990s, the use of ketamine as a party drug was becoming more common in many countries, and so it was being viewed in a more hostile way. Soon after, Russia banned ketamine, making it impossible for Krupitsky to continue his research.

Krupitsky's work wasn't forgotten. Professor Celia Morgan, a psychologist at Exeter University, became interested in ketamine when she wrote her PhD on its harm to the brain. As she learned more, and having met Krupitsky, she began to see ketamine's paradoxical quality; it not only has the potential for addiction but can also be used to treat addiction.

Recently, Morgan published the results of her groundbreaking study, "Ketamine for Reduction of Alcohol Relapse (KARE)." The funding came from the UK Medical Research Council, not only the most prestigious source, but also the hardest to obtain.[23]

The study subjects had moderate to severe alcohol dependence. They were randomized into two groups: three intravenous ketamine sessions or the placebo, three intravenous saline sessions.

Then the two groups were split again. Half took an educational course about alcohol, and half had a new kind of short four-week mindfulness type of focused therapy to help them think differently and to learn to deal with the challenges of not drinking and their urges to drink.

The two best outcomes were both the groups who were given ketamine. And the best outcome of all was ketamine plus the new short, focused therapy. For the ketamine and therapy group, the outcomes were very exciting; over the following six months, they stayed abstinent for more than 80 percent of the time.[24]

The National Institute for Health and Care Research (NIHR), the NHS's research funding body, together with Awakn Life Sciences (who I work with), has just funded a larger replication study. If it turns out to have as beneficial results as the original study, it will likely mean NHS approval.

So far, we only know that these results last for six months. And we don't know what the optimal regime will be. Would two or four KARE treatments (one treatment is one injection plus therapy) be as good or better? However, as there are already clinics in the UK, U.S. and some other countries offering ketamine with or without psychotherapy, there will soon be more data. In the UK, Awakn clinics are offering treatment courses based on Morgan's KARE Protocol in London and are setting up clinics in the United States too.

## TREATING ALCOHOL DEPENDENCE WITH PSILOCYBIN

Psilocybin wasn't tested on alcohol dependence in the 1960s, but the LSD data on this did look promising. The key modern studies have been done by Professor Michael Bogenschutz at the University of New Mexico (he is now at NYU Langone Center for Psychedelic Medicine).

The first, on ten people, was done on the psychedelic therapy model: twelve weekly sessions of therapy and one or two sessions of psilocybin, the first in week four.[25]

After a single treatment, subjects went from having an average of 25 percent heavy drinking days to 7 percent. After 36 weeks, people had still massively reduced their drinking.

Bogenschutz's much larger follow-up trial came out in August 2022.[26] It was a randomized double-blind trial on 93 people. Half were given psilocybin, and half were given an antihistamine as an active placebo (although, and this is a problem with psychedelics trials, people could tell the difference). By the end of eight months, nearly half those given psilocybin had stopped drinking.

Bogenschutz's next trial will be larger, with over 200 people.

What's interesting is that, like in the depression studies using psilocybin, the more intense the reported experience the better the gains people made in their drinking habits.

Just like other studies in this chapter, it seems psilocybin can reframe people's attitude to alcohol. Mary Beth Orr, 69, who took part in the study, reported that three years later she only has the occasional glass of wine. "It's not that I monitor my drinking," she said, "it's just that I don't think about it, which is the glorious part for me."[27]

The research also suggests that people with alcohol disorders may need higher doses of psilocybin, though we aren't yet sure why.

## TREATING ALCOHOL DEPENDENCE WITH MDMA

As a child psychiatrist in Taunton in the UK, Dr. Ben Sessa worked with traumatized children. Too often he saw that as soon as they turned 16 they would leave the care system and end up living on the street.

When someone has PTSD, a reminder of a traumatic event will evoke the exact emotion that occurred at the time, so they relive the original traumatizing event, again and again. It's now known that people often take drugs and drink to suppress trauma, and this is what these young people were likely doing. As they got older, they went from being looked after and treated as victims to being treated as criminals.

As you'll read in Chapter 8, there's good evidence that MDMA alongside psychotherapy can help people with PTSD. Sessa's theory was that by using MDMA to treat the trauma underpinning the drinking, these young people would then be able to stop. He set up the world's first addictions study using MDMA therapy, and the Bristol Imperial MDMA in Alcoholism (BIMA) study.

The study recruited alcoholics from the local Bristol treatment service and put them on an eight-week course of standard therapy for alcohol dependence plus two MDMA sessions. This was an open trial, so the subjects knew what they were getting.

After nine months, the majority were abstinent. Only a quarter of those who'd had MDMA were still drinking more than 14 units a week. There was an interesting result for those who were still drinking: some seemed to have reset their drinking behavior, so they were able to do it moderately, as in the psilocybin study.[28]

Compare this to the previous year, when people at the same treatment service received the standard eight-week therapy course. By three months, two-thirds were back to drinking more than 14 units a week.[29]

The study established that MDMA didn't have any adverse health effects, which is important as alcoholics have a higher risk of health complications. It also asked the question: Do people get depressed after taking MDMA? It's long been thought that MDMA depletes the brain of serotonin. But the research showed that MDMA therapy didn't lead to negative mood in these quite vulnerable people.

## IS IBOGAINE A SAFE DRUG
## TO TREAT ADDICTION?

Ibogaine is widely used in underground treatment centers around the world for the treatment of heroin and more recently methamphetamine addiction.

It's legal in Gabon, where it originates, as well as Brazil, Costa Rica, Mexico, the Netherlands and South Africa. In New Zealand, where it's an approved medicine, addiction treatment groups have been using it in the treatment of heroin addiction for a number of years. A published report of their treatments shows good outcomes with minimal adverse effects.[30] However, the data is from open (non-double-blind) interventions, which may bias in favor of positive findings.

There is a major downside to ibogaine treatment. As we saw in Chapter 3, it can affect the heart rhythm, which can lead to major complications including cardiac arrest. This is probably why deaths have been reported in ibogaine treatment sessions. The heart complication is particularly problematic because ibogaine is usually given during opioid withdrawal, when the physiological strains on the heart are at their greatest.

It may be possible to minimize cardiac problems by using ibogaine after a successful standard detox, although this would extend the time and costs of a treatment package. Recent work from Professor Paul Glue's team in New Zealand used noribogaine, the active metabolite of

> ibogaine. They showed that the impact on the heart is dose related, so using lower doses with careful cardiac monitoring seems a sensible way forward.[31]
>
> There is also research interest in some derivatives of ibogaine that are free of cardiac problems.

# DO PSYCHEDELICS WORK FOR OTHER ADDICTIONS?

As we saw above, the early trial using LSD for heroin had promising results. At Imperial, at the time of writing, we have funding for a pilot trial of heroin addiction using psilocybin to see if it's safe and tolerated.

There has also been a trial of ketamine for cocaine addiction.[32] It had a weaker effect than when used for smoking and alcohol. However, as cocaine is very addictive, directly targeting the drive/motive key brain system, and we have no substitute medication, it may be worth further research.

## ARE PSYCHEDELICS BEING USED FOR ADDICTION?

At present, the only psychedelic drug medicine available in most countries outside of studies is ketamine. It's showing good results at specialist centers, although it's not yet widely used by addiction services.

**How Long Can Effects Last?** We don't have studies looking at either ketamine or classic psychedelics for more than a year, so we can't be certain. But it looks as if the latter have the potential to change the brain in longer-lasting and more profound ways. For example, they seem to give people more insight into why they do the unwanted behavior, as well as a bigger perspective on their purpose in life.

A lot of their value seems to come from people having a big experience while under the drug. For example, people in the smoking cessation study (above) described seeing their life at the big-picture level and found a sense of inner connectedness that went beyond smoking.

**What Kind of Therapy Works Best Alongside Ketamine?** As yet, this is unclear. Do we use the usual types of therapy, or try a new approach like the shorter one in KARE? Not only might it be more likely to succeed, but it's quicker for the person, and cheaper for the treatment provider, which will hopefully eventually be the NHS and other national health services.

Patients also seem to like the fact that KARE therapy is an individual therapy. As one of the Experts by Experience patients who helped design the trial said, "Being in groups isn't for me, and that's why this appeals really." Another said they liked the fact it was new: "It's different, it's a new type of therapy, which is good for those of us who have been around the block and tried everything…" And therapists like it too:

> In my position (as a recovery worker), it's just such a struggle getting people to engage. We run groups and no one comes but this seems to provide some solution to that. Maybe the ketamine helps people engage with the therapy a bit better and that, that's a really good thing…

Also, it can work like a cure that allows people a more normal life:

> I don't like feeling like I am dependent on something else, medication every day or whatever, so this feels like it would be an improvement on that.

The psychedelic experience may result in more enduring neuroplasticity too. This will help patients to take advantage of the integration therapy sessions to make any necessary changes (see Chapter 8 for more). For alcohol dependence, it gives the person time to learn skills they need, such as monitoring when they lose control, or knowing their trigger points, or dealing with stress in other ways. It may be that after a psychedelic experience, the support of a group such as AA becomes more meaningful to people.

If we want to develop new forms of therapy to go alongside these new medicines, it may be hard to get addiction services to change. Therapists can often be ex-addicts who are wedded to the therapy that got them well, and there's a whole industry of companies and institutions who currently provide therapy.

### Do Patients Have to Give Up Drinking Altogether?

This is also unclear. In both the psilocybin studies and the BIMA MDMA study, it was found that a proportion of people could reset to being able to drink without losing control. But at present we can't identify these patients before they try out alcohol again, so we still encourage everyone to stay abstinent.

### Which Drug Is Best for Treating Addiction?

Currently, there aren't enough studies to show this. However, in the future perhaps we will be able to offer different drugs to target the different reasons that people are addicted. We could offer MDMA, for example, if the reason someone drinks is related to childhood or life trauma. And psilocybin if you started drinking because you enjoyed it and then found you couldn't stop. And as psychedelics work on a brain-wide level, in the future perhaps we will be able to offer them for non-substance addictions, such as gambling, internet or porn. These are questions that need to be addressed in the research.

# Chapter 8

# MDMA: PTSD, TRAUMA AND RELATIONSHIPS

**IN 2009,** I was sacked from my position as the drugs czar—aka chair of the UK Advisory Council on the Misuse of Drugs (ACMD). The catalyst was that I spoke out about the real harm of drugs, specifically about the harm of ecstasy. At the time, I did not imagine that MDMA would one day be a medicine.

In the UK, MDMA had been a controlled drug since 1977, long before it became popular or a problem. It was banned not on its own but as part of a group of amphetamines. The illegality of MDMA and its bad press was no barrier to its rapid rise in popularity in the late eighties and early nineties. It created a whole new way of partying: acid house, raving, the so-called Second Summer of Love and electronic dance music. After the 1994 Criminal Justice and Public Order Act, the government's heavy-handed legislation to ban raves, partying moved to legal, licensed clubs and superclubs.

As had happened with LSD in the 1960s, both the media and the government were treating MDMA as one of the most dangerous drugs. Both major political parties took a "tough" stance on young people taking ecstasy because they knew it was a vote winner. Some young people were, very sadly, dying after taking ecstasy (see Chapter 12 for why). But these deaths were massively overreported compared to deaths from other drugs: one study showed how in 2008 the press reported every death that involved taking ecstasy but only 2 percent of deaths from alcohol poisoning and 9 percent of deaths due to heroin (for more on the risks of MDMA, see Chapter 12).[1]

My firing, though stressful for me, did have one benefit—it started a public discussion around the harm of drugs. But there was still a severe lack of good science about MDMA's real effects. You could trace this back to the 1980s when it was banned in the U.S. The U.S. then led a successful campaign at the UN for MDMA to globally be put in Schedule 1, which states that a drug has no medical use.[2]

Now, 25 years later, in Australia MDMA-assisted therapy has been approved as a treatment for PTSD that has failed to respond to conventional treatments. And it is very likely that it will soon be available as a treatment in the U.S. and Canada.[3, 4]

The driver of this improbable U-turn is the activist and campaigner Rick Doblin and the nonprofit Multidisciplinary Association for Psychedelic Studies (MAPS), which he founded in California in 1986. Doblin's stated long-term goals are mass mental health and global spirituality. And for nearly 40 years, MAPS has been lobbying for legal change, and supporting and sponsoring research into the safety and therapeutic applications of psychedelics.

The organization's major focus is on MDMA-assisted therapy being licensed to treat PTSD. PTSD is both hard to live
with and hard to treat. As a psychiatrist, I found PTSD cases
the most disheartening. Even the best talking therapies and
medications only work around a third of the time.[5]

The licensing of MDMA as a treatment has been expedited
by the enormous need of U.S. veterans returning from the wars
in the Middle East. The prevalence of PTSD in the general
population is between 1 percent and 12 percent.[6] However,
among U.S. military veterans, it can be up to 29 percent,
depending on the conflict the veterans experienced.[7] Vets with
PTSD are at high risk of suicide. One estimate says that
almost four times as many U.S. servicemen and women killed
themselves in the ten years after 9/11 than were killed in
action.[8]

So far, MAPS has conducted six successful phase 2 trials,
and have recently completed the second of the two phase 3
trials that are necessary for the treatment to become a regis-
tered medicine. In the first phase 3 trial, results were way
beyond the threshold needed for a substance to become a
medicine: 88 percent of participants had a "clinically significant reduction in symptom severity." And 67 percent no
longer qualified as having a diagnosis of PTSD, compared to
32 percent in the placebo group.[9, 10]

As results were positive, MAPS has submitted a dossier to
the FDA asking for approval. And as the FDA has declared
MDMA-assisted psychotherapy for PTSD a "breakthrough
therapy," it is likely to be granted, first in the U.S. and then also
in Canada and Israel.[11]

\*      \*      \*

My group at Imperial College has played a small part in this land-mark achievement. In 2012, we did the first ever imaging study to show how MDMA worked in the brain. During the study, we also imaged emotions during memory recall, important in PTSD. It was a somewhat extraordinary study because it was televised.[12]

I had wanted do imaging into the brain effects of MDMA for some time. At this time ecstasy was an established party drug, being used by hundreds of thousands of people in the UK every weekend. And yet very little was known about how MDMA works in the brain to produce its effects.[13]

It was known that MDMA worked on the serotonin system, and that huge doses were neurotoxic in animals (see Chapter 12 for more), but not how MDMA changed the brain. MAPS had started their research program, despite opposition, and I was intrigued by their preliminary studies, which indicated that MDMA-assisted therapy might be useful for PTSD. This made sense: MDMA's three Es, empathy, energy and eupho-ria, seemed uniquely suited to psychotherapy.

I started to look for funding, but I was turned down by the usual institutions. I was told by one off the record that it was considered too much of a reputational risk for them to fund research on an illegal drug. It became clear to me that there was a very real stigma around researching these drugs; scien-tists and institutions working in this field were at risk of being seen as encouraging lawbreaking, despite the research itself being not only legal but essential. For a scientist, this kind of negative reputation could be career ending. Or at least job ending, as I'd found out.

Regulations about obtaining and then keeping illegal drugs in a lab, and the cost of purchasing them from a permitted source, made research both difficult and expensive, in both time and costs. This seemed ironic: every Friday and Saturday night, there were people buying ecstasy pills in their local town for the price of a few drinks.

Then I was approached by the public service broadcaster Channel 4, who told me they were interested in filming cutting-edge scientific research on an illegal drug in humans, specifically a live broadcast of people taking cocaine. I declined as the brain research on cocaine had already been done. They were disappointed but, a month later, came back and asked, "What Class A drug would you be prepared to study live on TV?" Immediately, I replied "E." I suggested we use their funds to do a proper MDMA imaging study. At Imperial College, we'd recently finished our first imaging study looking into what psilocybin did in the brain (see Chapter 5) and we were blown away by the results, in particular that it turned off, rather than turned on, high-level regions of the brain. I told Channel 4 that it would be fascinating to see if MDMA would act differently. They said yes.

It took a year for Imperial College to give permission for the study, and required me to write hundreds of emails and to have a face-to-face meeting with the Home Office.

## THE 1970S: MDMA AND THERAPY

MDMA was first used for its therapeutic properties in the U.S. in the 1970s. The psychiatrist and psychotherapist Professor Torsten Passie found that from the late seventies until it was banned in 1985, a few dozen therapists, mainly in California, were using MDMA in a psychotherapeutic setting.[14] And MAPS estimates that around 500,000 doses were used in therapy in North America during this time.[15]

This wide use with very few reported harms was a helpful part of the evidence that allowed us to convince regulators to approve the ongoing clinical trials of MDMA.

Unlike the classic psychedelics, MDMA rarely gives hallucinations or other sensory changes. At the time, the general consensus among therapists was that MDMA was useful because the patient's feelings of anxiety and fear would be dampened and they'd feel a general sense of well-being that would deepen their relationship with their therapist. Claudio Naranjo, a Chilean psychiatrist, used MDMA with more than 30 patients. He said it produced "artificial sanity, a temporary anesthesia of the neurotic self."[16] Therapists used MDMA to enhance individual therapy but also in groups and couples, to help people feel empathy toward each other, trust each other and communicate better, so breaking down resentments. Naranjo found it useful for "clearing away the garbage" that had built up in a relationship.[17]

In the US, Leo Zeff, who'd previously used LSD in therapy, gave MDMA to around 5,000 people in group sessions, and trained more than 150 therapists to use it.[18]

His weekend group sessions started on a Friday night with a sharing circle, where each person would talk about what was on their mind. They had to sign an agreement not to leave without permission, not to be sexual or aggressive, and to stop doing anything they were asked to. On the Saturday, after taking MDMA, people would put on headphones and eyeshades and be encouraged to stay lying down. The instructions were: "If you don't know what to do and your mind wanders, then listen to the music. If you go into heavy judgments against yourself, then listen to the music." Zeff wrote that sometimes, "people like to get up and do some hugging and then we set them right back down." There was a communal meal on Saturday night, then Sundays were another sharing circle.[19]

The word spread that MDMA allowed people to be less defensive and share their feelings, and helped them talk about difficult issues and memories without becoming overwhelmed by the difficult emotions that usually come with them. Psychiatrist and professor Joseph J. Downing said: "Depending on the material contained in the unconscious, the patient will deal with any situation, from childhood traumas, to long-felt adult insecurities, to deeply repressed emotions."[20]

Ann Shulgin, a Californian therapist (and the wife of Alexander "Sasha" Shulgin, who discovered MDMA's properties), started using MDMA to help people with PTSD. "I don't know how many people have come up to us, and said, 'You've changed my life,'" she said.[21]

## HOW AND WHY MDMA WAS BANNED IN THE U.S.

The therapists using MDMA were understandably cautious about publicizing their work, given what had happened to LSD once it started to be used outside the lab and therapy room. But inevitably for a drug that gives energy and euphoria as well as empathy—and is legal—by the early eighties, MDMA was being used as a clubbing and party drug.

In the U.S., it soon became clear that the DEA were intending to ban MDMA. A group of therapists, activists and researchers formed to fight the ban. Their aim was to conduct research to provide the DEA with evidence that MDMA did have therapeutic benefits. If they could prove medical use, MDMA couldn't be put in the most restrictive category of illegal drug, Schedule 1. Despite a three-year battle and two judicial decisions that went against the DEA, eventually MDMA was put into Schedule 1, where it has stayed ever since. The activists regrouped under the leadership of Rick Doblin and formed MAPS. As with psychedelics, the U.S. ban led to a UN global ban, which spelled the end of therapists around the world being able to use MDMA in therapy.[22]

That is, except in places where MDMA went underground. In Switzerland, from 1988 to 1993 therapists were allowed to apply for a special license to practice. But when it was banned in 1993, some continued to use it illegally. In Zurich, German psychotherapist Dr. Friederike Fischer held weekend group sessions for clients, using MDMA alongside LSD, until she was arrested in 2009.[23]

Fischer's therapy was psycholytic rather than psychedelic. This involves repeated low doses of drugs as opposed to psychedelic therapy, which involves large doses given just a few times. Fischer invited patients in her practice who she considered "stuck" to attend weekends at her house, where the group would all take MDMA, followed by LSD or another psychedelic. Over the years, she treated nearly 100 people, who attended an average of 25 sessions each.[24]

Fischer was arrested, reported by a patient who blamed Fischer for her marriage breaking up. During the trial, Fischer told the judge: "For me psychedelics like MDMA and LSD are not drugs. They are psycho-integrative substances that have been used for thousands of years. (It) is not like getting drunk. The clients are in a clear state of elevated consciousness in which they can carry out psychotherapeutic work." She was fined 2,000 Swiss francs ($1,500) and given a 16-month suspended sentence with a following probation period of two years.[25]

## HOW DOES MDMA HELP PEOPLE PROCESS TRAUMA?

Dr. Michael Mithoefer, a US psychiatrist and lead researcher for MAPS, calls PTSD "a cluster of symptoms that happens after traumatic events." "It is striking that, regardless of the kind of trauma—war trauma, physical assault, sexual abuse, childhood abuse—many of the kinds of symptoms people have are basically the same kinds of symptoms."[26]

People with PTSD constantly reexperience their trauma and the emotions that come with it. Sufferers are easily startled and quick to anger. They might have flashbacks and intrusive memories, dark thoughts, panic attacks, nightmares and other sleep problems. They might avoid things that remind them of the trauma, and so disconnect from emotions and from friends and family. They are more likely to suffer from anxiety, depression, cognitive and memory problems and addictions, and are at high risk of suicide.

### CLASSIC PSYCHEDELICS AND PTSD

MDMA is not the only drug being trialed to treat PTSD. When we did our first psilocybin depression study in 2013 (see Chapter 5), the only predictor of negative outcome

170

was anxiety during the trip. I was concerned that people with an anxiety disorder such as PTSD might not do well on classic psychedelics, so we didn't pursue this.

However, in the past ten years I've heard a lot of anecdotal evidence about classic psychedelics and PTSD, mostly from veterans who have been to ayahuasca retreats in the Amazon or Spain or psilocybin sessions in the Netherlands (see a personal account of this on page 209). I have now changed my mind. As soldiers are usually traumatized in a group, perhaps the group aspect of these retreats is healing as well as the treatment. It might even be more powerful than solo therapy, particularly if the retreat is with other veterans.[27, 28]

Compass Pathways are now doing a study of psilocybin and PTSD.[29]

There is also some preliminary evidence for ketamine where a small study of six treatments found powerful benefits. Clearly a larger trial is now warranted.[30]

## WHAT IS MDMA-ASSISTED THERAPY FOR PTSD?

When Dr. Mithoefer came across MDMA, he had a private practice specializing in PTSD with his wife, Annie Mithoefer, a psychiatric nurse. He was looking for more effective treatments. "It really came out of clinical need," he has said. "I wasn't against them [existing psychiatric medicines] in some global way, but I was very disappointed in how effective they were, and it was very clear we needed something better."[31]

The Mithoefers led two of the early MAPS studies, starting in 2004, including one for veterans as well as firefighters and police officers.[32, 33] They are now responsible for both training

therapists and developing the MDMA-assisted therapy protocol that will be used alongside MDMA when it is licensed.[34]

Mithoefer describes MDMA-assisted therapy as being unlike
other psychiatric medicines, which aim to decrease symptoms.
Instead, MDMA therapy targets the psychological causes of
those symptoms. "So, in that process, they might feel more
symptoms before they feel better," he says. "It can be part of the
healing process if they are supported at working through it."

He says it's important to warn people that the experience
may not be pleasant. "Patients have said, 'I can't believe why
they call it ecstasy.' It can be challenging, difficult and painful.
People are going where they haven't been able to go before.
Processing trauma is challenging with and without MDMA.
It just might be possible with MDMA."[35]

When it's used as a treatment, MDMA is usually given in
two or three doses over 16 to 18 weeks, with psychotherapy
sessions before the MDMA sessions to prepare the patient,
and after for integration.

"At first glance, MDMA-assisted psychotherapy looks very
different from any conventional treatment: participants lying
on a bed, sometimes with eye shades and headphones, listening to music, with male and female therapists sitting on either
side for at least eight hours," says Mithoefer.

Like therapy using psilocybin, MDMA sessions have a "nondirective" approach: that is, the patient takes the lead on what
they want to talk about. This allows them to bring the trauma
up in their own time. "The trauma always comes up, and we
think it is preferable to allow it to come up at whatever time
and in whatever way it does so spontaneously for each individual," says Mithoefer.[36]

Research shows that this nondirective approach is as effective for even the hardest-to-treat types of PTSD, such as dissociative PTSD, where the person feels disconnected from their emotions and reality. It also works for trauma with different causes, whether it's childhood or adult, multiple events or a single event.[37]

The results of the therapy last too, even up to 74 months after the end of treatment.[38]

This is how people who've had the therapy describe it:

I keep getting the message from the medicine, "trust me." When I try to think, it doesn't work out, but when I just let the waves of fear and anxiety come up it feels like the medicine is going in and getting them, bringing them up, and then they dissipate.

Without the study I don't think I could have ever dug down deep, I was so afraid of the fear.

Maybe one of the things the drug does is let your mind relax and get out of the way because the mind is so protective about the injury.[39]

## KEY ELEMENTS OF MDMA-ASSISTED THERAPY

* Screening. MDMA does increase pulse rate and blood pressure, so it isn't suitable for people with significant vascular or heart disease.

* Preparation sessions. These take place with the two co-therapists, usually male and female, who'll be present throughout all the sessions.

* MDMA sessions. People usually take the MDMA in the morning of the session. Sometimes people will wear their eyeshades and headphones and listen to music, alternating with times they'll speak to the therapists. The therapists use open-ended questions, inviting the participant to express their thoughts and feelings and tracking the participant's emotional state and process.

* Integration sessions. These happen the morning after the MDMA sessions, then several times over the next few weeks. The therapists will invite the participant to discuss their experience and help them make sense of it.

## ECSTASY ON TV

We finally began our MDMA study for the Channel 4 program in 2011. Each of the 25 participants had two brain scans, a week apart. For one scan they were given a capsule of MDMA, and for the other a placebo. The study was double-blind; neither the participants nor the researchers knew which one they would be given first. However, given ecstasy's effects, afterward nearly everyone correctly guessed which they'd been given.

*Drugs Live: The Ecstasy Trial* went out on September 26 and 27, 2012. It was, almost certainly, the first serious scientific study funded by a television company. And I was told that at the time it was the most downloaded program ever in the history of Channel 4, exceptional for a scientific documentary. More importantly, it was the first whole-brain fMRI study of MDMA's effects and the most detailed analysis of the brain actions of MDMA ever performed.[40]

The results were as surprising as the source of funding. We discovered that the major effects of MDMA were happening in very different regions of the brain from psilocybin. Rather than affecting the cortex, the highest-level part of the brain, it worked in the emotional center (the limbic system). There was an increase in communication between the amygdala, the brain's alarm system, and the hippocampus, which coordinates memories. These two areas of the brain are intimately involved in stress and the laying down of traumatic memories. People with PTSD have a reduction in communication between these two areas of the brain.[41]

A second change was a decrease in activity in the limbic or emotional system. The more intense the reported effects of the drug, the more this happened.[42]

Thirdly, MDMA also reduced activation in a part of the brain called the insula, which becomes active when you experience fear or anxiety.[43]

The three brain effects are exactly what you'd want to see in a drug to treat PTSD. We think that reducing activity of the emotional and fear centers (the limbic system and insula) and increasing connections between the alarm center (amygdala) and the memory center (hippocampus) allow the person to go back to their traumatic memory, reprocess it, and engage with therapy to help them move forward.

## REMAKING MEMORIES

We did some specific tests on what MDMA does to memory too. Before their scans, we asked each participant to write down some of their most important memories, six good and six bad.[44]

They didn't share the content of the memories with us but gave each one a label that we could use as shorthand. For example, a good memory might be "Honeymoon in Paris," and a bad one, "Told my friend is dead."

While being scanned, each participant was asked to hold each memory in their mind for 20 seconds, so we could see the response. After the scan, they were asked to score the vividness and emotional strength of each of their memories on a 0–100 scale. To allow us to compare the placebo and MDMA, the participant recalled half the memories (three good, three bad) during the placebo scan and half during the MDMA scan.

We found, as predicted, that under MDMA people rated their good memories as more vivid and more positive than under placebo. They rated their bad memories as equally vivid under MDMA, but the negative weight of the memories was lower.

These memory effects suggest a reason why MDMA might work for PTSD.

In the figure opposite, you can see that the factual part of the memory is encoded in a different part of the brain than the emotional part of the memory. The emotional memory is usually the most distressing aspect of PTSD. For example, it's very disabling if, every time you hear a door slam, your brain interprets it as you are being fired on like when you were in a war zone and activates the same emotions and fear responses as it did then.[45]

It's almost impossible to eliminate the factual memory. But we can eliminate the emotional one. Regular therapy involves having people remember the trauma, so they relive the trau-

Figure 10: The location of trauma memories in the brain and where MDMA works to reduce emotional reactivity and facilitate recovery

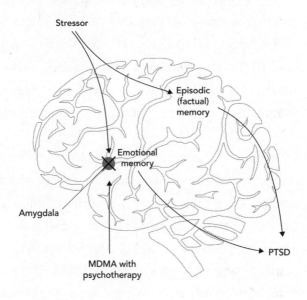

Source: Adapted from Nutt DJ, de Wit H, 2021, Putting the MD Back into MDMA, *Nature Medicine* 27: 950–951, ISSN: 1078-8956.

matic memory alongside the therapist, who keeps them feeling safe until the emotion extinguishes (wears off). But this is too hard for a lot of people with PTSD.

MDMA, by reducing the emotional response, allows people to stay in therapy until the emotion is extinguished. Afterward, they can talk about the event without their emotions resurfacing as they used to do and overwhelming them.

## MDMA AND YOUR BRAIN CHEMISTRY

Like SSRIs, MDMA dampens emotional responses (see Chapter 5 on depression). But as we showed with the memory test, unlike SSRIs MDMA dampens negative responses only. It pushes the emotional circuit in the direction of positivity.[46]

While psilocybin, SSRIs and MDMA all work on the serotonin system, they do so in very different ways. Classic psychedelics work by stimulating the 2A receptors directly and so mimicking some of serotonin's action. SSRIs raise levels of serotonin by blocking reuptake of serotonin.[47] And MDMA directly increases levels of serotonin by stimulating its release.[48]

MDMA doesn't only work on serotonin receptors. Like other amphetamines, it releases noradrenaline, which focuses attention and gives energy, and dopamine, which also gives energy as well as the motivation to dance and talk. In theory, dopamine release should make MDMA addictive; this is what drives cocaine's addictive power. But MDMA isn't very addictive, which is probably down to its much greater serotonin effects (see Chapter 12).

However, some of our more recent research suggests there must be other pharmacologies of MDMA that we don't yet understand. There are other, similar molecules that release neurotransmitters in a similar way to MDMA but don't produce the same level of empathy as MDMA.

We do know MDMA can stimulate the release of oxytocin, which may be how it enhances trust. Oxytocin is critical for mother–baby bonding, allowing the letdown of milk. Its role in adults is less clear, but it's often called the love hormone because it enhances couple and group bonding, facilitating trust and warmth. Some experiments have shown that oxytocin administration via a nasal spray improves social empathy.[49]

We did look to see if the effects of MDMA in our imaging study were correlated with oxytocin levels in the blood, but they didn't seem to be. However, the release of oxytocin in the brain may still play a part in helping the person trust and bond with their loved ones—or their therapist if they're in therapy.

There is a theory from work in rodents that oxytocin promotes neuroplasticity.[50] If this occurs in humans too, it could explain how MDMA helps people reframe the impact of their old fear-related memories and so, like psychedelics, enables people to move on from them.

## ECSTASY WITHOUT THERAPY

It may be that MDMA can work, to some extent, without therapy. In our study, out of all the participants it was J, a priest, whose story was the most noteworthy.

J wanted to take part in the study because she'd heard MDMA had the potential to be a treatment for trauma, which she thought was the cause of much of the homelessness and alcoholism issues in her parish.

She also had PTSD after suffering a serious assault, compounded by other personal traumas, which we found out after her MDMA session. She'd had treatment with EMDR and felt herself recovered.

She described how ecstasy felt. "When it hits you, you can feel your heart racing, feels like your body is melting." After this initial difficult stage, she found it easier. "Later on when it's not so intense, you get loads of happy memories coming back. The feeling was happiness, all the times in my life I've had those happy feelings coming back."

Then she described how she felt about her bad memories. "When I reached back for the bad memories, they did not seem bad. In fact they seemed to have been fatalistic necessities for the occurrence of the later good events I had to recall.

"The thing with PTSD is, you can't reach those emotions, you can't express them. You are left with anger and feeling disconnected from the people who mean the most to you...I thought, that is the first time I've shown emotion about it." She said that reconnecting to her memory allowed her to process it at a deeper level than her previous EMDR treatment. "My brain is free to pick good memories, not locked into blocking it."

A few days later, she had perfect recall of her assault for the first time. "I had a full memory recall in chronological order." But she saw it differently from before. She was even able to laugh at it. "I was thinking, no wonder I've got PTSD!" Despite J being given MDMA as part of an experimental study and not as a treatment, she found it helped her recovery. She said that MDMA had "clearly done something quite profound."

However, this isn't a recommendation for taking MDMA outside a clinical setting without therapeutic help. For some people this could bring up trauma rather than treat it.

## MDMA FOR ANXIETY

Considering MDMA's calming effect on emotional centers of the brain, it makes sense that it may help people with kinds of anxiety disorder other than PTSD.

* Anxiety in adults with life-threatening illness. So far, study data looks promising. It's interesting that both classic psychedelics and MDMA appear to help these patients, even if the drugs have different mechanisms (see Chapter 10 for more).[51, 52]

* Social anxiety in autistic adults. Autism isn't a single disorder, so it's unlikely that this therapy will be helpful for everyone. A lot of people with autism struggle with extreme social anxiety. The treatment aim isn't to change the person or make them less neurodiverse but to help them deal with their anxiety when they meet other people, especially strangers, allowing them to engage with the world and suffer less.[53]

# THE FUTURE OF RESEARCH

There is work going on to develop group therapy for PTSD in veterans,[54] and a study has started at Yale to look at the brains of people who have PTSD on MDMA.[55]

As well as PTSD, MAPS has sponsored or supported multiple studies for other uses of MDMA, including in couples therapy, for anxiety for adults with life-threatening illness and for social anxiety in autistic adults (see previous page). There are studies planned into anorexia and binge-eating disorder as well as social anxiety disorder, among others.[56]

At Imperial College and Bristol University, we used MAPS-manufactured MDMA in our research into alcohol dependence (see Chapter 7 for more). The results showed that MDMA is very effective at helping people with this, possibly as addiction problems are often due to underlying trauma.[57]

There is still a lot to do, though, in particular in finding out how MDMA changes the brains of people with different conditions.

In 2000 a BBC *Horizon* program told the story of Tim Lawrence, a film stuntman who, at the age of 34, was diagnosed with Parkinson's disease, a condition caused by loss of dopamine and usually associated with older people. Tim spent every day either frozen or moving and shaking uncontrollably due to a combination of the illness and the medicine he was prescribed, L-DOPA. He discovered that MDMA helped stop this: "Standing in the club with thumping music claiming the air, I was suddenly aware that I was totally still. I felt and looked completely normal. No big deal for you, perhaps, but, for me, it was a revelation," he said.[58] And the program showed him exercising in the gym, doing swallow dives and backflips.[59]

Later animal studies supported Tim's experience. Research then focused on trying to find a legal alternative. However, since the UK Psychoactive Substances Act, any substance that is similar to MDMA is illegal. At Imperial College we identified several potential MDMA analogues, but the Parkinson's research laboratory didn't have the Schedule 1 license that would have allowed them to study them. Other groups have come up against the same issues.[60]

It will be fascinating to see how this area of research progresses in the next few years, and it will be especially heart-warming to see patients finally being able to access a treatment that works where others have failed.

# Chapter 9

# END OF LIFE, MYSTICAL EXPERIENCES AND FINDING GOD

**IN HIS LAST** hours, bedridden and dying from cancer of the larynx, Aldous Huxley wrote a note to his wife, Laura, asking her to inject him with 100mcg of LSD. In his novel *Island*, published the previous year, 1962, Huxley had created a society where the inhabitants took "moksha-medicine"—aka psilocybin mushrooms—to expand their consciousness, improve their quality of life, make a better society and also to ease them into death.

Laura later wrote in a letter to Aldous's elder brother Julian and his wife, Juliette: "Suddenly he had accepted the fact of death; he had taken this moksha-medicine in which he believed. He was doing what he had written in *Island*, and I had the feeling that he was interested and relieved and quiet."

Around an hour later, "the expression of his face was beginning to look as it did every time that he had the moksha-medicine, when this immense expression of complete bliss and love would come over him." An hour later she injected him with another 100mcg and, a few hours after that, he died.

Huxley had, Laura described in the same letter, "the most serene, the most beautiful death." She wrote: "Aldous was, I think (and certainly I am) appalled at the fact that what he wrote in *Island* was not taken seriously. It was treated as a work of science fiction, when it was not fiction because each one of the ways of living he described in *Island* was not a product of his fantasy, but something that had been tried in one place or another and some of them in our own everyday life. If the way Aldous Huxley died were known, it might awaken people to the awareness that not only this, but many other facts described in *Island* are possible here and now."

Laura then asked a question that's still relevant today: "Now, is this way of dying to remain our, and our only relief and consolation, or should others also benefit from it?"[1]

Evidence is building that psychedelics may not only help people die a peaceful death but face death when given a terminal diagnosis too. And, more than that, that the spiritual dimension of the psychedelic experience may help people who are facing all kinds of difficulties to find meaning in life, and even to find God.

## A MATTER OF LIFE AND DEATH

Most of us are very frightened of dying. When people are given a terminal diagnosis, and even before that, when they are told they have an illness that may be terminal, not surprisingly, a good number become anxious and/or depressed.

Antidepressants do not always work for this group of people. There's minimal research but the little there is suggests that they may not be any better than placebos.[2] This is possibly because these people are suffering from something more existential than being depressed and/or anxious. The symptoms have been summed up under the label demoralization syndrome[3,4] and can include a loss of meaning or purpose in life, low mood, feeling helpless, disheartened, trapped, or like a failure, not being able to function or cope, and having thoughts of suicide or wanting to die sooner rather than later.

Currently, the way we help people die (and sometimes help them get through the weeks leading up to death) is to stultify or intoxicate them with opiates so that they can't think or feel. They may be pain-free but their body is constipated, their brain is numbed and they often don't have the ability to speak cogently before they die.

When I did my clinical training at Guy's Hospital Medical School in London, over 50 years ago, doctors gave people in pain who were dying from cancer a medicine called Brompton's mixture: a cocktail of heroin (or morphine) and cocaine.

I remember one lecture where the excellent physician, clinical pharmacologist and teacher Professor John Trounce was teaching us how to help people at the end of their life. I had read Laura Huxley's description of giving Huxley LSD on his deathbed and I put my hand up, and asked, "Should we consider giving dying people LSD?" Professor Trounce said, "I don't know anything about that. And we don't talk about that. And in any case, it's illegal." I didn't say what I was thinking (which is unlike me). But it was "Hang on, morphine and cocaine are illegal too."

As opiates and antidepressants aren't the answer, as doctors we still have a huge gap in our ability to help people at the end stage of their life. There's now growing research to suggest that Huxley's choice was right: that psychedelics may fill that gap.

The first scientific study into psychedelics to treat severely ill people was published in 1964, the year after Huxley's death.[5] Fifty people were given either a single 100mcg dose or two oral opioids. People who were given the LSD reported not only having less fear of death but also better pain control in the short term (for more on pain treatment, see Chapter 10). From the 1960s to the mid-seventies the researchers did two more studies that showed good results; although the participants knew what they were getting, the studies did give a good indication that LSD can give you a better death.

This thread of research came to life again in 2011. There have now been four trials on people with life-threatening illnesses including advanced cancer, using both LSD and psilocybin.[6, 7, 8, 9] Together, they suggest that psychedelic therapy can help a lot of the symptoms suffered by people in this situation: anxiety and depression, but also suicidal thoughts and the other symptoms of demoralization; plus they also increase spiritual well-being. The studies haven't revealed any adverse effects, suggesting that the therapy is safe even for physically fragile people.

Newer studies are also looking at a wider range of illness that's not terminal, for example AIDS, as well as psychedelic-assisted therapy in group settings, a way of making psychedelic treatment more cost-effective.[10]

## WHY DO PSYCHEDELICS HELP PEOPLE TO A BETTER DEATH?

The antidepressant actions of psychedelics are key. Immediately after a trip, people often report a more positive mood that can last days and even months. This is probably due to the same mechanism as in depression (see Chapter 5), where the psychedelic switches off the negative thinking circuit.

But there is something else going on too. Taking psychedelics is a reliable way to alter consciousness to produce a spiritual, mystical or transpersonal experience, all names for gaining a sense of there being something beyond or bigger than the person, of being pulled out of your everyday mode of thinking. The British philosopher Dr. Peter Sjöstedt-Hughes, author of *Modes of Sentience: Psychedelics, Metaphysics, Panpsychism*,[11, 12] says there are two more elements to this profound change in consciousness: there is the out-of-body experience or ego dissolution, where the barriers between the self and the world appear to break down, and there is the metaphysical, where how we used to explain the world breaks down, so that our perception of life and the universe is fundamentally changed.

This can lead to profound changes in the way people view themselves and the world. A mystical experience can be a sense of being loved, either by others or by themselves. This feeling of love may have a conventional religious flavor, so the person feels God's love. But it often extends beyond this into a sense of being part of an interconnected community of humans, nature and sometimes everything. Many people express having a sense or knowledge of love as a universal emotion that connects all people and indeed all things. In some, this might be the Christian sense of the unconditional and omnipresent love that God has for his people, whereas in others it may be more a sense of one-ness with nature or the universe.

One reason people are frightened of dying is that they see it as the end. The power of these drugs to alleviate suffering is in part due to their ability to help the patient see that it isn't. Key to this is the experience of ego dissolution, fusing or becoming one with, or more connected to, the world or universe. People may feel as if they are floating out of their body, even looking down on themselves. In more extreme cases they can feel their body seeming to spread into space, their body identity and integrity disintegrating. Yet they are still the same person, just in a different form.

Logically, we know the atoms that make up each of us will continue to exist and will inevitably be incorporated into new life at some point. Psychedelics allow people to engage with that thought emotionally. Psychedelics are comforting, because they can bring the person to realize that that they are not going to die in their entirety, that death is a transition rather than an end. As the Christian death ceremony has had it since long before the discovery of the science of the cycle of life, "dust to dust, ashes to ashes."

## ELEMENTS OF EGO DISSOLUTION

These are from the Ego-Dissolution Inventory,[13] a self-report scale designed to measure the extent of ego dissolution after psychedelics.

* I experienced a dissolution of my "self" or ego.
* I felt at one with the universe.

* I felt a sense of union with others.
* I experienced a decrease in my sense of self-importance.
* I experienced a disintegration of my "self" or ego.
* I felt far less absorbed by my own issues and concerns.
* I lost all sense of ego.
* All notation of self and identity dissolved away.

## WHAT IS A MYSTICAL EXPERIENCE?

For millennia, human beings have used different methods to alter consciousness and bring on a mystical experience. Methods include sensory deprivation, fasting, sleep deprivation, meditation and ritual drumming, chanting, singing and dancing, sweat lodges and, of course, natural and now manufactured psychedelics.

In our brain imaging studies a significant number of the participants, both healthy volunteers and patients, reported some kind of mystical or spiritual experience. For example, in our first depression study about a quarter of the patients reported this after their high-dose psilocybin treatment, with two specifically saying they'd experienced God.

In that study, we measured the size of each patient's positive psychological experience. This is the "peak" experience that can be religious, mystical or spiritual. We found that the higher the peak experience, the better the antidepressant response. The two people who had mild trips didn't seem to do as well.[14]

It's not clear whether it's the amount of the drug in the

brain that's important (more being likely to produce a more profound effect) or the spiritual experience itself. People with good outcomes commonly reported that they had a greater sense of "connectedness" with nature and others after the trip. One person said, "[After the dose] when I went outside, everything was very bright and colorful, and it felt different. I noticed things I didn't notice usually, the leaves on the trees and the birds, small details."[15]

Another said, "A veil dropped from my eyes, things were suddenly clear, glowing, bright. I looked at plants and felt their beauty. I can still look at my orchids and experience that: that is the one thing that has really lasted."

For some, these improvements in perception stayed. For example, one person said six months later, "Things look different even now. I would look over at the park and it would be so green, a type of green I'd never experienced before. Being among the trees was incredible, like experiencing them for the first time, so vibrant, so alive."

For others, the experience during the trip was so intensely felt that they struggled to put it into words. Later, one person described it like this: "I was everybody, unity, one life with six billion faces, I was the one asking for love and giving love, I was swimming in the sea, and the sea was me." Another said: "Like Google Earth, I had zoomed out, I was absolutely connected to myself, to every living thing, to the universe."

Two patients said they felt the presence of God. One said, "Not God in some dogmatic way, a God-like archetype within your psyche, that is real and within you. I know this exists, I directly experienced it. I was suddenly taken in a rapture, and I was floating in midair, with my eyes wide open and my mouth open, completely in a state of awe and ecstasy. It's a very powerful message to take away."

The other said: "Then I felt the presence of God: I have always thought that he was a man because of the way I was raised, reading the Bible, but it felt like a female energy." Another patient described a female "ancient being" who, although they did not refer to it as God, was "omnipotent and unconditionally loving."

Some saw religious imagery during the dose, such as temples and Hindu gods. One said that at the peak of the experience, "I was Shiva, dancing." Others described a powerful feeling of "love" as a supernatural force.

## WHERE DO MYSTICAL EXPERIENCES COME FROM?

The religious view is that the psychedelic—or the fasting, meditation, prayer or sleep deprivation—is the channel that allows the mystical experiences to come from outside the person. In other words, the experiences come from a god, and the brain can be tuned to receive them.

There's an apt quote that's usually ascribed to Pierre Teilhard de Chardin, a twentieth-century French philosopher, paleontologist and Jesuit priest whose theory of cosmology included evolution. (For this, his books were banned by the Roman Catholic Church.) It goes: "We are not human beings having a spiritual experience. We are spiritual beings having a human experience."[16]

Ethnobotanist Terence McKenna, who died in 2000, described psilocybin mushrooms as extraterrestrials in a symbiotic relationship with human beings. "I think that it is possible that certain of these compounds could be 'seeded genes' injected into the planetary ecology eons ago by an automated space-probe arriving here from a civilization somewhere else in the galaxy," he wrote.[17] In his version, the mystical experience is a type of communication from aliens.

As a neuroscientist, I believe that mystical experiences on psychedelics emerge from the brain, and that this is the case for other forms of altered consciousness too, such as dreams, mania and delusions in schizophrenia. But I have always been fascinated by why humans exist, why we have religious belief systems and where they come from. As a child, I remember a discussion with a classmate over who made the world. She said, "It was God." I asked, "So who created God, then?" I suppose I'm a sort of reductionist, in that I believe that subjective experiences come from the brain. In terms of human psychedelic experiences of other dimensions and existences, my current thinking is more agnostic than atheist. It's more in line with that of Sir Martin Rees, the former president of the Royal Society and Astronomer Royal, whose research on the origins of the universe discussed the implausibility of existence being a random event. His book *Just Six Numbers* tells the incredible story of the six variables, imprinted in the Big Bang. Each had to be near perfect in their magnitude, otherwise there would be no stars, no life—including us human beings. In 2011, he was criticized when he accepted the $1.2 million Templeton Prize, because the prize has a spiritual agenda and was created for "Progress in Religion." He told the *Guardian*: "Doing science made me realize that even the simplest things are hard to understand and that makes me suspicious of people who believe they've got anything more than an incomplete and metaphorical understanding of any deep aspect of reality."[18]

Studying the nature of human consciousness requires a similar level of humility. Over the course of my career, I have repeatedly been surprised, mesmerized and awestruck as some of the complex and intricate workings of the brain have been discovered. There is no doubt that psychedelics are a brilliant tool for helping us discover what happens in the brain when someone has a spiritual experience. They provide a reliable way to induce, in a controlled scientific setting, these special states of mind that have inspired thought leaders and influenced human history. Psychedelics are certainly a much easier and sometimes safer way to achieve major insights into the special capacity of our brains than days of starvation or sleep deprivation, or years of meditation, or any other historic way to achieve transcendence.

But then how to explain each individual's personal experience? The great German psychiatrist and philosopher Karl Jaspers said there were two elements to this: form and content. The form of a psychedelic trip can now be explained in terms of altered brain connectivity. However, the content, i.e., what you see, feel and think, can't yet—and may never—be explained by imaging and indeed science. We can make suppositions using a person's history or psychology, but we can't disprove someone's belief that they have encountered entities in other dimensions during a trip.

This work suggests that extreme experiences such as near death experiences (NDEs) (see box on page 197) come from changes in the way the brain works, not from changes outside the brain. In fact, the most likely explanation of NDEs is a lack of oxygen to the brain; free divers who go for long periods without oxygen also report going into mystical-type states.

## MEETING THE ENTITIES—AYAHUASCA AND DMT

At Imperial, we have studied one of the psychedelics that's been most associated with having a mystical experience, DMT (see page 31), the psychedelic in ayahuasca. When people take ayahuasca or DMT, they often describe losing their sense of self, entering a different dimension, feeling a sense of oneness with the universe, and even seeing or sensing extraterrestrial beings or entities. Another common experience during a DMT trip is being transported into another dimension, where the entities are.

Some describe it as like passing through a "wormhole" into hyperspace or through a door into another universe. Others say that it's entering a higher dimension of consciousness where a powerful ambience of love exudes from the entities or even just from the new dimension/space. This new dimension or universe is felt as being in ways more real, more vivid and more desirable than the current one.

At Imperial, we couldn't study ayahuasca itself: the variable composition of the plant material and its preparation makes giving a standardized research dose very difficult (although it has been done), so instead we used the active ingredient DMT.

We got some advice on setting up our studies at Imperial College from Professor Rick Strassman. He's a psychiatrist and Buddhist who pioneered DMT research from 1990 to 1995 and reported his findings in the book *DMT: The Spirit Molecule*.[19] Working at the University of New Mexico, he gave nearly 60 people around 400 doses of DMT. However, his research was beset with problems of regulations and the resulting costs and, eventually, he had to give it up.

Strassman confirmed that Western scientists can get similar

effects from DMT in the lab as shamans do using ayahuasca in the Amazon. And he also confirmed that DMT was safe to administer.

The challenge of doing research with DMT is that it's quickly broken down in the liver and blood and the effects last less than ten minutes, not long enough to get good imaging results. Ayahuasca isn't broken down as fast as chemical DMT because it's brewed with two plants: one that contains the DMT, and one that slows the breakdown process. But we knew it'd be impossible to get permission to give research subjects a second, also untested chemical on top of DMT.

The study lead, neuroscientist Dr. Chris Timmermann, worked out a way to administer DMT that would give us long enough to scan: two intravenous injections about fifteen minutes apart. We tried it with a low dose of DMT (7mg) and found that it produced marked visual changes, which got bigger at a higher (14mg) dose. The highest dose (20mg) produced the typical kind of trip that people describe from taking ayahuasca.

Once the subjects were out of the scanner and the trip had ended, we asked them to draw their visual experiences too. Several who got the highest dose drew images of shapes resembling people or entities.[20]

In our second DMT study, we scanned people after giving each person the same: two doses of 20mg and placebo intravenously. The idea was to detect whether, when people see entities, anything specific is happening in their brain. At the time of this writing, we have published the psychological and imaging data but analysis of the brain location of entity experiences is complicated and ongoing.[21]

## CAN DMT CAUSE A NEAR-DEATH EXPERIENCE (NDE)?

Near-death experiences (NDEs) are a brain state that happens when someone comes close to death. People who've survived them report spiritual experiences that, at first glance, seem very similar to what's reported with DMT, such as feelings of going out of their body, going into another world, inner peace and communicating with entities.

Mammals, including humans, produce tiny amounts of DMT in our brains, particularly in the pineal gland.[22, 23] And because these two states of altered consciousness seem to have so many similarities, researchers developed a hypothesis that, at the point of death, we may have a spike in our own inner DMT. However, that hypothesis was undermined by a detailed scientific analysis conducted by medicinal chemist Professor David Nichols.[24]

He pointed out, among other arguments, that the amount of DMT that's produced by the pineal gland is many times smaller than the amount needed to produce any kind of psychedelic experience.

At Imperial, we collaborated with the Coma Science Group in Liege, Belgium, who research NDEs. People who'd taken DMT took the same questionnaire about their experiences as people who'd had an NDE.

Questions included: "Did scenes from your past come back to you?" and "Did you see, or feel surrounded by, a brilliant light?"

When we compared their results, we found a considerable overlap in symptoms, especially when it came to ego dissolution and mystical-type experiences. The NDE group had more religious and transcendental/out-of-body experiences, but otherwise the scale of the subjective reports was largely the same.[25]

## Figure 11: Comparison of the changes in different aspects of consciousness produced by DMT (in black) and near death experiences (in white)

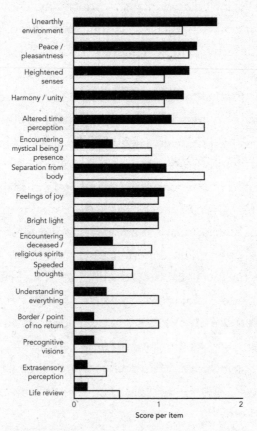

Source: Adapted from Timmermann C, Roseman L, Williams L, Erritzoe D, Martial C, Cassol H, Laureys S, Nutt D, Carhart-Harris R et al., 2018, DMT Models the Near-Death Experience, *Frontiers in Psychology* 9, ISSN: 1664–1078.

## BRAIN CHANGES AND THE MYSTICAL EXPERIENCE

Not only does having a mystical type of experience make it more likely that the person will feel mentally and emotionally better, but research has shown that having one has been linked with people changing their personality as well as their values and views of the world.

During our first depression study, a small study of twenty people with depression, we also tested personality traits three months after treatment. The table below shows the effect of psychedelics on the "Big 5" personality traits, considered to be the key traits that define personality. As you can see, people reported feeling more open, extrovert and conscientious and less neurotic. Interestingly, the latter two changes are also reported for people taking conventional antidepressants, but changes in extroversion and openness are specific to psychedelics.[26]

| Neuroticism | How much a person experiences the world as distressing, threatening and unsafe. Emotional instability. | Down |
|---|---|---|
| Extroversion | How much a person's interests and energies are directed toward the outer world of people and things rather than the inner world of subjective experience. | Up |
| Conscientiousness | How much someone is organized, goal directed, able to follow rules and delay gratification. | Up |

| Openness | How much a person is open to new experiences of all kinds, intellectual and cultural. | Up |
|---|---|---|
| Agreeableness | How much a person tends to act in a cooperative, unselfish way. | Unchanged |

### The brain becomes more open

Openness is interesting because the personality trait gives a snapshot of what's happening inside the brain.

An example of increased openness after taking psychedelics is a color-blind surgeon who reported that he could see the red of blood for the first time. Psychedelics don't alter the visual signal from the retina because the person will still have the defect, fewer "color-sensing" cones. But I think psychedelics facilitate new brain links between sensory input and our concepts of colors.

This is the bigger picture. As we read in Chapter 4, psychedelics open the brain in many directions; it becomes more receptive not only to different ways of hearing and seeing but also to different ways of thinking and being. This new openness of the brain, to re-see and reinterpret the world, happens on many levels.

We also think openness might be the beginning of or a necessary preparation for having a spiritual experience. To misquote the French chemist Louis Pasteur, chance only favors the mind that is prepared.

### Feeling more connected

After a psychedelic experience, people report feeling more connected to themselves, to other people and to the wider world. This is important because feelings of disconnection are associated with mental and emotional suffering. As bell hooks wrote, "Rarely, if ever, are any of us healed in isolation. Healing is an act of communion."

Dr. Rosalind Watts, when she was at Imperial College, formalized this increased sense of connectedness into the Watts Connectedness Scale (WCS).[27] A survey of over 1,200 people, as well as questions given to participants in the psilocybin versus antidepressant trial, showed that all three elements—connectedness to self, others and to the world—increased after taking psychedelics.

Some of our healthy volunteers who took part in our scanning studies reported feeling more connected with nature after psilocybin or LSD sessions, despite the drugs being taken in the rather unsettling environment of a scanner.

The cofounder of Extinction Rebellion, Dr. Gail Bradbrook, has spoken about how her experiences with psychedelics gave her connection to "the oneness." "You are just here for some time, and you have in some way a role to play to contribute to this magnificence of life and that is somehow your purpose." She went further, saying: "We are not here to be in the oneness, there will be lots of oneness when we're dead. How can we use these medicines to really connect with each other?"[28]

### The experience of ego dissolution

In itself, ego dissolution might be enough to make someone think more mystically about themselves. The person might be amazed that their brain can do more remarkable and different things than they ever thought possible. And the sense of the body disintegrating and interconnecting with the world beyond may bring its own mystical feelings.

When we did an online questionnaire of psychedelic experiences,[29] it showed that the more a person experienced ego-dissolution under a psychedelic, the greater their mystical experience. This suggests that either the two may arise from the same brain changes, or that they may indeed be the same brain phenomenon expressed in different ways.

Our work with both psilocybin and LSD has shown—almost certainly—that ego dissolution comes from turning off a part of the brain called the posterior cingulate cortex (PCC) (see Chapter 4). The job of the PCC is to integrate input from the senses, especially sight, plus the sense of place and time. It is the master controller of "normal" consciousness that tells you where and when you are.

When the PCC stops speaking to the rest of the brain, the brain becomes confused as to where its body is in space and time, which leads to a breakdown of the sense of self.

One of the brain areas that the PCC becomes detached from is the anterior cingulate cortex (ACC). As experiences of anxiety are largely coordinated in the ACC, this may explain why such a profound "out-of-body" experience is generally not anxiety provoking.

The shutdown of the PCC as the cause of ego dissolution was confirmed by a case report of an operation to remove a brain tumor. The surgeons began by mapping the tumor using electrical stimulation, which has to be done while the patient is awake. The stimulation turns off healthy brain tissue but not the tumor itself. When the PCC was stimulated, the patient stopped speaking and became unresponsive to questions. After four seconds, he came out of this episode and described being "as if in a dream, outside the operating room." After the second time this happened, he said: "I was as in a dream, I was on the beach." And after the third, "I was as in a dream, I was surrounded by a white landscape."

Interestingly, the surgeons also reported that the patient had better mood and well-being for weeks after the operation.[30]

## INSIDE THE MYSTICAL BRAIN

There is one bigger change in the brain on psychedelics that feeds into all the personality, sensing and brain changes we've talked about so far: the default mode network (DMN) going offline. The DMN is the brain system in charge of defining your self or your ego. Our imaging studies showed that mystical experiences were strongly associated with a breakdown of the DMN.

As described in Chapter 4, the brain works by creating models about the world. What we think we "see" or "hear" is not what enters our eyes and ears. Rather it is a combination or fusion of sensory inputs with our brain's internal predictions. The brain makes predictions about the outside world, analyzes and interprets information coming in, compares the two, refines its prediction by linking it with relevant memories, then creates plans on which to act. This process allows us to act and respond very much faster in real time than if we had to scan and interrogate the environment on a millisecond-by-millisecond basis.

Usually, these predictions are pretty good and save the brain a huge amount of work. Sometimes, though, they are wrong, and then you can see the brain at work. For example, your brain expects to see work people in a work context. If you've ever bumped into a colleague at, for example, the supermarket or an airport, you may have found you were slower to recognize them.

The brain does the same model-generating process to produce thoughts, imaginings and ideas.

You even do this while you're asleep. The stories your brain makes up—your dreams—reflect your mental state—for example, if you're worried or stressed or sad—as well as your concerns. People with PTSD dream of the trauma, recently bereaved people dream of their dead loved ones and some accountants even dream of lists of numbers!

Some of your brain treats these stories as real. We know this because heart rate and blood pressure both increase during nightmares, and eye movements track dream images. The brain even switches off the centers that are responsible for limb movements so that we can't act out our dreams. This active suppression breaks down in a condition called REM Behavior Disorder, often a prelude to Parkinson's. The person then acts out their dreams, fighting or trying to protect themselves from perceived attacks, hitting out and often hurting themselves or their bed partner. The fact that this suppression exists and usually works shows that somehow the brain "knows" our dreams are perceived as real by some parts of the brain, including the movement-generating areas.

It seems possible that, in its attempt to make sense of the altered inputs that happen while on psychedelics, the brain makes up spiritual models. And if you are a believer, it frames them as religious experiences.

## SERPENTS VS ALIENS

Why do Amazonian people report seeing jaguars and serpents when they take ayahuasca, but Western people report aliens, entities or God? After taking psilocybin and LSD, a few of our subjects described being in the presence of God. They said God was a feeling, a sense of immense and meaningful presence or being, often in a splendor of bright white light.

Sometimes Western ayahuasca users are disappointed when they don't see the jaguars, serpents and other beasts that the locals do.

In his book *PiHKAL*, Alexander Shulgin describes what happened when he introduced the writer and psychedelic therapist Claudio Naranjo to Richard Evans Schultes, one of the pioneers of the botany of psychedelic plants. Their conversation, which took place in the 1970s, shows how the same drug can have very different effects, depending on the background and expectations of the person taking it.[31]

Claudio: "What do you think of the jaguars?"

Richard: "What jaguars?"

A short silence.

Claudio: "Are you personally familiar with authentic *Banisteriopsis caapi*?" [one of the ingredients in ayahuasca]

Richard: "I was the person who assigned it its name."

Claudio: "Have you ever taken the plant decoction itself?"

Richard: "Perhaps fifteen times."

Claudio: "And never jaguars?"

Although there hasn't been any specific scientific research on this topic, the most likely explanation is that the content of psychedelic experiences is culturally determined, just as dreams are.

Amazonian tribes that use ayahuasca learn from an early age about animals such as snakes and jaguars, which are central elements of their religion. When under the influence of ayahuasca these are therefore what they see. The tendency for psychedelics to produce visual hallucinations that look like snakes helps this along.

Western ayahuasca or DMT users often report traveling into space and seeing entities or aliens that also engage with and talk to them. These are likely to be products of our recent history of space exploration and the birth of science fiction.

Christian believers tend to see Christ, or the Christian God or some more abstract representation of God. They often see distant towns and palaces or cathedrals on hills or mountains. These are images that have been used in Western art for centuries, especially in religious art.

This is the human prediction-making machine—our brain—at work again, primed by our cultural expectations. We see what we expect to see, perhaps also what we wish to see.

## CAN WE FIND GOD IN THE BRAIN?

It seems unlikely that it could be possible to induce a religious belief by interfering with the brain of someone who doesn't want to believe. But for a believer or someone who's open to having faith in a god, the sense of transcendence that psychedelics give, or being in the presence of something or someone beyond oneself, becomes religious. Research shows that we can use psychedelics to help people seeking religious or spiritual enlightenment to get there faster and perhaps further too.

The Good Friday Experiment, which took place on that day in 1962, was an ambitious attempt to assess scientifically how a drug might induce or enhance a religious experience. It was one of the great experiments in the history of psychopharmacology.

Dr. Walter Pahnke, who led the experiment, was a Harvard physician and colleague of the Harvard psychedelic research team that included Timothy Leary and Dr. Richard Alpert (who later became the spiritual teacher Ram Dass). But Pahnke was also a church minister and so was interested in helping people achieve spiritual awareness of God.

The experiment took place in a small basement chapel, the Marsh Chapel, while the usual service was going on above. As part of his PhD research program, Pahnke gave two groups of trainee priests either a high (30mg) dose of psilocybin or, as an active placebo, nicotinic acid.[32]

Nine of the ten who received the psilocybin experienced profound religious feelings. Only one of the ten placebo-treated participants did. This 90 percent success rate led to the experiment being called "the Miracle of the Marsh Chapel." And those priests who experienced these profound religious insights maintained their beliefs for decades, if not forever.[33]

Researchers at Johns Hopkins University and NYU are doing a modern version of this experiment.[34] They recruited a number of religious leaders from several different faiths, including Judaism, Christianity and Islam, and gave each one two doses of psilocybin. Afterward they questioned them about their faith. The final results haven't been published, but early reports say most have become more spiritual and more confident in their religious beliefs afterward, just as in the Miracle at Marsh Chapel.

The Episcopal priest Hunt Priest said: "I had a very Pentecostal experience the first time—feeling my body, speaking in tongues. I felt a force of energy blow into my body. Now

I see that spiritual healing is real. I'd laid hands on people and prayed with people, but frankly, I was just touching and holding space. Now I know healing is real. It's a little embarrassing to admit it, but I was in my head all the time. This was in my body.

"I *did* experience the Holy Spirit. I'm clear about that. I felt like my journey at Hopkins was a second ordination."

Since then, he has set up a Christian society to organize psychedelic retreats for clergy called Ligare, from the Latin word meaning to bind. "Religion at its best," he says, "binds us to God."[35]

## CHANGING CONSCIOUSNESS

Some of the nondrug gateways to spiritual and religious experiences—meditation, for one—have been shown to change your physiology in similar ways to psychedelics to make your brain less rigid. I now suspect a lot of these approaches, such as sleep deprivation, chanting and dancing, work to suppress the DMN in a similar fashion to that produced by psychedelics. They affect the brain so we can feel mystical and "out-of-body" experiences, as well as strengthen our belief in a god if we are so inclined.

For most of us, psychedelic drugs are the easiest way to gain access to these feelings and insights because the changes they provoke in the brain are so effortless and reliable. They disrupt conventional patterns of seeing and experiencing the world that have been laid down and have become more and more ingrained since childhood. By turning off the control centers of the brain, they produce novel experiences that many people make sense of as mystical or even being in touch with God. Or, at the very least, give us a sense of wonder and awe.

After being diagnosed with anxiety and depression after serving in Afghanistan, the British soldier Keith Abraham traveled to the Peruvian jungle to take ayahuasca.

I served in the Parachute Regiment for close to nine years, from 2004 to 2012, including in Iraq and Afghanistan.

In Afghanistan, the fighting was particularly intense, and we were at the heart of it. We were under attack and fighting every day. Some of my friends died, and a lot had life-changing injuries. I was present for a lot of those incidents and I took bodies off the battlefield while under fire. That went on for four months out of the seven months I was there.

Although this was what I joined up for and what we trained for, it was also pretty traumatic. The battalion lost ten people on that tour, five of them from our small Forward Operating Base.

Even before I left Afghanistan, I felt conflicted about what had happened. It felt like a waste of time, that I'd lost so many friends for no reason. I didn't see what we had achieved, apart from killing people, even though that's what we were there to do. I came home thinking, "this is probably going to have some impact on me."

It was a year until I experienced any symptoms of trauma. I was having a nice time in a bar with my then partner. We were talking, I'm not sure about what but it wasn't

Afghanistan. And I became aware my eyes were watering, that I was physically crying but I wasn't emotionally crying. It was pretty disturbing. I thought, "I'm holding on to something so tightly that it's leaking out of me." It was clear to both of us that I was troubled.

I asked for help from my GP, who gave me antidepressants and sent me to a counselor.

The pills numbed me completely, which I didn't like at all, so I stopped taking them. And the counselor was a young woman. I couldn't tell her what I'd seen, it felt as if I was traumatizing her. We weren't the right match.

I was clearly still in trouble. By this time I'd started sweating excessively, day and night, from my hands, face, feet and back. And I was losing my hair so fast that in the morning, my pillow would be covered in it.

My behavior had changed too. I didn't have the vocabulary or emotional intelligence to say what I wanted to say to my partner. I'd become frustrated, she'd become frustrated, we'd both get upset, we'd both cry—but my tears would be hysterical. We kept going around in circles. When our relationship finally broke down, I think it traumatized me all over again.

I went back for more psychotherapy with a different therapist and different antidepressants, but still nothing worked.

By this time, I'd left the military and I was working in the City. Within a short space of time, two friends messaged me, separately, and told me they'd been to Peru and had had significant experiences on ayahuasca, and they thought it was worth me trying.

I took a two-week holiday from work and flew to Peru. I would be spending ten days alone in a simple wooden hut, called a tambo, deep in the jungle, an hour's walk from the nearest village. The jungle was so dense, you could only see the sky if you looked straight up. The hut had no running water or electricity but there was a river 20 yards away, where I could bathe. I had to carry in all my water and food for the week.

The people who owned the hut walked me there. They showed me the trees where I could pick fruit, bananas and a red fruit called a pomarrosa. And they introduced me to Juan, whose cacao farm was 500 meters away, and who'd be looking out for me. Then they left.

I spent a lot of time just sitting and being in the jungle. I was free to go anywhere. The jungle itself was medicine. There were little black monkeys who would rob my bananas— I'd come back to my hut, and they'd all be eaten. I wondered why they bothered to steal my bananas when they could easily get their own. When I needed company, I spent time with Juan. A couple of evenings, Juan killed and cooked a chicken for dinner.

I was told to expect the shaman on two particular days, late in the afternoon. And to eat only breakfast on those days.

On the first appointed afternoon, I was sitting outside the hut and Don Aquilino appeared on the path by the river. He was in his fifties or maybe sixties, the elder of the village. He wore jeans and a T-shirt, not the traditional shamanic garb you might have seen in pictures. That's because he's a mestizo, which means he's originally of Spanish, not Shipibo (the indigenous people) descent. He was carrying a small Coca-Cola bottle filled with a brown fluid. He didn't speak any English, and I only have a small amount of Spanish, so we communicated as best we could.

Juan came to look after me physically, and Don Aquilino was in charge of my spiritual well-being. The evening was the same both the times I took ayahuasca. We waited and chatted until nightfall. Then the shaman poured me a cup of ayahuasca. He told me to drink it, then to lie down and close my eyes.

He began to sing the *icaro*s, the traditional chants and songs. Some are repetitive chants, some are clearly songs with verses and choruses, some are just sounds. But they are magical, the vehicle upon which I was taken on the journey. Even if you hadn't drunk ayahuasca, they'd still be beautiful. Sometimes they are fast and intense, hard to listen to, but they are designed to take you through the difficult part of your journey and make you purge, either physically or emotionally. You might feel a sense of nausea, then you'll reach a peak and either vomit or purge in some way, perhaps diarrhea, burping, crying, coughing, retching, some way to release the tension. I cried. It felt like a relief. It is as if the medicine is trying to get you to purge and when it's done, then the *icaro* changes, becomes soft and calming and relaxing.

I had many visions during the two experiences. In one, I realized I'd come into a different realm, a very old-fashioned Victorian classroom with a chalkboard. In front of me was an old woman, and the medicine was talking to me through her manifestation. She showed me my behavior patterns and choices, and how I was the cause of my own suffering.

She took me back to landmark moments of my life when I had reacted with violence or anger or frustration. And then she played the part of me in those interactions, so I could witness her responding to the people involved with compassion, forgiveness, patience and kindness. She was teaching me how to do this.

And then she said, "Now you are going to live those opportunities again, and I want you to respond with compassion and forgiveness, patience and kindness, toward them and yourself, and with gratitude."

Each landmark moment was a test. If I managed to respond in the way she was asking, then I'd pass that test and she'd give me another test, another landmark moment. When I failed, I'd have to do the test again. She was so kind and patient. She'd say, "Do it again, remember to be patient, to understand the person shouting at you, try to find compassion, to forgive them, to see they are in pain."

None of the tests was about the army. One was a time when a drunk guy stumbled into me and spilled his food on me, then dropped it, then got angry with me. In the real-life situation, I'd ended up headbutting him. In the ayahuasca test, I'd headbutt him again and fail. Again and again, until I didn't. Over what felt like lifetimes, I learned how to extract myself from each of those situations.

That whole night was one continuous spiritual teaching. At some point, I realized it had ended, that I was so tired. The shaman said, "You're finished," and I rolled over and went to sleep there. When I woke up, the shaman said, "You worked hard, well done."

*     *     *

I left that jungle knowing that I still had work to do on the relationships, although I had started on that side of my life. But I knew that the trauma of my combat experiences was healed.

At home, I decided to help my friends from the army access psychedelics too, then I realized that all veterans deserved access to these treatments. I started the UK branch of the U.S. charity Heroic Hearts Project (https://heroic heartsproject.org/our-story/). We have taken a group of veterans to Peru, and plan to take groups to the Netherlands for psilocybin treatment too. We are also setting up an observational research study into the potential effect of psilocybin on traumatic brain injuries in veterans (see Chapter 8) in partnership with Imperial College, and we are starting a study to look at the effects of ayahuasca on PTSD with psychiatrist Dr. Simon Ruffell. I hope that very soon everyone who needs psychedelics will be able to access these treatments, because they are life-changing. More information at heroicheartsuk.com.

# Chapter 10

# CLASSIC PSYCHEDELICS: ANXIETY, PAIN, EATING DISORDERS, ADHD AND OCD

**AFTER OUR FIRST** successful trial of psilocybin therapy in depression at Imperial College, we began to consider other disorders that psychedelics might be helpful for. Based on our thinking that they work by disrupting repetitive ruminative processes in the brain and breaking down maladaptive thought habits, we began to research anorexia nervosa and OCD. Both of these conditions are characterized by fixed negative thought patterns.

Other research groups are studying psychedelics for conditions that would benefit from these drugs' ability to increase neuroplasticity. They are looking at conditions where the brain needs growth and rewiring, including stroke and traumatic brain injury.

While the conditions that involve fixed thoughts require therapy, the conditions that require brain growth also require rehabilitation. It's interesting how adaptable psychedelics can be;

it appears their value is that they allow the brain to relearn and change in whatever way is needed.

This chapter lays out the research in some key treatment areas. Each is marked out of ten for the quality of evidence so far. Most of the scores are low because almost all the research is at very early stages; however, there are now a number of commercial pharmaceutical companies investing in studies, which makes it increasingly likely that some of these treatments will be available within the next ten years.

## ANXIETY

Evidence rating: 3/10

Now that psilocybin is getting closer to being licensed for depression, attention is moving to anxiety. This makes sense: SSRIs, which like classic psychedelics act on the serotonin system, are prescribed for anxiety. And psilocybin was used as a medicine for anxiety in some countries in the 1960s, under the name Indocybin.

A recent survey showed that having a psychedelic experience (outside the lab) was associated with having a lower level of anxiety symptoms.[1] Some U.S. studies have shown good results in reducing anxiety in terminally ill patients. And preclinical studies have shown repeated doses of LSD to have antianxiety effects, albeit only in mice that were already stressed.[2]

There haven't been any trials in humans on anxiety alone. Because people who feel more anxious during a trip tend not to get such a good outcome, it may be that to treat anxiety the psychedelic needs to be given alongside an established anxiety medication. A Swiss study, not yet published, gave a high dose of LSD to people who were taking SSRIs for anxiety but were not fully recovered. Preliminary reports suggest that LSD produced a trip in most of them, despite the SSRIs, and that it reduced their anxiety.[3]

# CLUSTER OR EPISODIC HEADACHES

Evidence rating: 9/10

Attacks of cluster headaches have been described as being more painful than childbirth, and even as more painful than being shot. The sharp, burning or piercing pain is so hard to bear that sufferers have reportedly tried to escape it by rocking, pacing, banging their heads, running out of the house or shoving their face in ice-cold water. Another name for cluster headaches is suicide headaches, because people with this condition have high rates of attempting it. "The brutality and severity of the pain...wreaks havoc on relationships, family life, employment and friendships," says Ainslie Course, a director of the cluster headaches charity ClusterBusters UK.[4]

Cluster headaches are more common in men and often come on in the middle of the night. It's not certain what causes the headaches; it's thought they may be a type of migraine. Existing treatments are often not effective.

It's been known for some time that (usually) micro- or midi-sub-hallucinogenic doses of psychedelics every few days can not only treat attacks but lead to long-lasting remission.[5]

This is slightly paradoxical; psilocybin can give some people headaches, so much so that we are cautious about accepting headache sufferers into trials.

The treatment may work via psychedelics' neuroplasticity effects (see Chapter 5 for more). But it cannot yet be prescribed as it's currently illegal.

A few years ago, I was sent three patients with cluster headaches who'd failed to respond to all established treatments, including being on oxygen all night.

I suggested to each one that they might want to try psilocybin or magic mushrooms.

One said he wouldn't dare try it because it was illegal, and he was worried that if he was caught he might lose his pension. One said he'd try it, went out and picked mushrooms, took them once a week, and stopped having headaches. The third said he wanted to fund a study on psilocybin in the UK, but at the time the main charity, ClusterBusters UK (mentioned above), declined to help, citing illegality. More recently, ClusterBusters UK have begun campaigning for a change in the law.[6] It is still very easy for people to misinterpret the law and assume it is illegal to use psychedelics for research when in fact it is legal.

Research is now under way too. The first clinical trial, a Danish study, gave ten patients three low-to-moderate doses of psilocybin (COMP360, Compass Pathways' proprietary synthetic version) once a week for three weeks. It showed that the frequency of attacks was reduced by 30 percent. One patient had complete remission for three weeks.[7]

There's also promising research on a non-psychedelic version of LSD called 2-Bromo-LSD.[8]

Given the challenges of getting psilocybin, some people have turned to alternatives, specifically LSA (D-lysergic acid amide), which has similar effects to LSD but can be extracted from the seeds of some plants in the morning glory family. N.B.: Although seeds that contain LSA are legal, as some are seeds of common garden plants, extracting LSA is illegal.[9, 10]

# CHRONIC PAIN AND PAIN SYNDROMES

Evidence rating: 4/10

Ten years ago, I became interested in psychedelics for pain when a patient with a mental health disorder shared his history with me. As a young teenager, he'd had severe chronic neuropathic pain from a cycling injury. This seriously disrupted his schooling and he could hardly walk. He couldn't find a treatment that helped. On his fifteenth birthday, he was taken to a club by friends and one of them gave him LSD. It dramatically lifted his pain. And although the pain persisted, he told me that it was nothing like as bad as it was before.

There are many different forms of pain syndromes, including what this patient had, neuropathic pain. This also happens when someone loses a limb and it still hurts; in that case it can also be called phantom limb pain. There is also fibromyalgia, a condition where the patient has pain all over their body as well as extreme tiredness; we have started a brain imaging study of the effect of psilocybin on this.

The current thinking is that psychedelics might work to reduce pain in a few different ways. Firstly, animal models suggest they are anti-inflammatory. Secondly, just as with depression, pain can become embodied into the brain, so people can't stop thinking about and feeling it. The pain gets uploaded into parts of the brain where it wasn't previously, then keeps being recycled. This process is beyond normal volitional control.

Imaging shows that people in pain have an overactive anterior cingulate cortex (ACC). This circuit, one of the higher parts of the brain, has the job of integrating motivation, emotions and memories. We know that psychedelics disrupt the ACC in the normal brain, so this may be a key mechanism. Psychedelics may disrupt pain circuits in the same way as they disrupt other out-of-control brain thinking loops.

Another interesting angle is that mindfulness training is becoming widely used as a method of pain control. As we saw in Chapter 6, psychedelics may work in the same way as meditation; so if psilocybin works, mindfulness could be facilitating the same process. And combining the two may be more powerful still.[11]

Lastly, an increase in neuroplasticity (the growth of new neurons and new connections between neurons) might allow the brain to reorganize and make new connections, overwriting pain networks.[12, 13]

## STROKE AND TRAUMATIC BRAIN INJURY

Evidence rating: 2/10

We tend to think of brain injury as a massive bump to the head. But in fact, it exists on a spectrum; it can be anything from a minor bump to the head that leaves you feeling confused and dazed for a few days to being unconscious for the rest of your life. And there's also a less obvious kind: mild traumatic brain injury or mTBI. This came to light when soldiers returning from war in the Middle East complained of problems with attention, concentration and brain fog. They didn't have obvious head injuries from being shot or hit, but they had clearly suffered some kind of injury to the brain.

Often, it turned out that their car, truck or tank had been blown up and this had shaken up the brain, damaging it in a similar way as happens in shaken baby syndrome. After this, the brain can function abnormally for months or even years. Sports injuries can also produce a form of mTBI. In sports, the amount of brain damage will depend on how many times you've been punched in the head, knocked out or headed the ball, plus the violence of each episode.

In mTBI, brain damage is diffuse—that is, it can occur all over the brain, causing a whole range of cognitive and emotional dysfunction. And so there isn't a good diagnostic method for mTBI, except people's history and symptoms.

These two types of mTBI have something in common with stroke, namely loss of neurons and their connections. Stroke is the colloquial term for acute brain injury caused by a blockade of blood vessels (a blood clot) or a bleed into the brain. These both deprive the brain of oxygen so that nerve cells die. The dead cells release toxins, which then kill other cells, and so the size of the damage progresses.

Until recently, attempts to reduce the impact of stroke have focused on getting oxygen back into the brain, for example, by using clot-buster drugs and sometimes surgery. There's a very short time window for this.

What's needed for all these kinds of brain injuries is a way of helping the brain heal by regrowing the neurons that are dead or dying. This is where psychedelics come in. Not only

Figure 12: Head traumas

| Strokes | Brain injury |
| --- | --- |
| • Bleed for aneurysm | • Diffuse micro-lesions across all brain = TBI (Traumatic Brain Injury) (e.g., from explosions) |
| • Clot blocking blood flow | • Focal from direct trauma (blows to head), e.g., frontal brain damage from hitting car windshield |
| | • Subcortical, e.g., Parkinson's disease in boxers (e.g., Muhammad Ali) |

for the recent finding that psychedelics can promote neuroplasticity, but also that, as well as promoting new neuronal processes, they might also reduce the neuro-inflammation produced when nerve cells are damaged and go on to do more damage to the brain.

Several researchers are investigating whether psychedelics might be helpful after a stroke. Two years ago, Algernon Pharmaceuticals approached me at Imperial College to do a DMT treatment study into people with stroke.

At first I was concerned that they wanted to put people who'd just had a stroke through a DMT trip, which would be unethical. They reassured me they planned to use a sub-psychedelic dose, based on a paper in rats that showed that DMT accelerated recovery.[14]

Algernon was granted ethical permission for the first trial in humans. My team worked out how to give a continuous, slow DMT infusion for six hours at a sub-psychedelic dose, which has been successfully trialed in healthy volunteers.

The other key question was when to give DMT. Giving it too soon after a stroke might be dangerous as it can elevate blood pressure. But give it too late, and you may miss the optimum window to help the brain to heal. We decided to give DMT a couple of weeks after a stroke. We are now doing the first safety trial in stroke patients.[15] The study also includes rehabilitation by limiting movement in the unaffected side of the body to force the brain to learn to move the affected side.

Based on this work, a colleague at Imperial College has started investigating using DMT to help recovery from TBI (Traumatic Brain Injury) in preclinical models.

## ALZHEIMER'S AND DEMENTIA

Evidence rating: 1/10

Alzheimer's and other types of dementia are currently a one-way street: there is no restorative treatment available that can regrow the brain. However, there have been case reports of psychedelics helping people become more alert, functioning and present in the world. Potential benefits would come from psychedelics' ability to promote neuroplasticity and help reduce neuroinflammation, as for brain injury (see above).[16]

The research in this area is very new. Eleusis Pharmaceuticals has trialed a range of doses of LSD in 48 older people to look at safety and effects on brain function. They were treated six times, every four days. These data suggest that midi-dosing is safe in older people. Further studies are needed in patients with cognitive impairment and Alzheimer's.[17]

# EATING DISORDERS

Evidence rating: 4/10

In anorexia nervosa, people—most of them young and female—severely constrain their eating to target a desired but very low body weight. The initial driver to restricting food is usually the person's (false) belief that they are fat, i.e., a distorted body image. On top of this, controlling weight can become a way of exerting control and gaining autonomy over their life. The person sometimes exercises excessively too. Over time the behavior seems to become embedded as a habitual resistance to eating, something like an addiction to thinness or perhaps to hunger. This continues even in the face of life-threatening weight loss and starvation.

Often, people with anorexia come to know they're harming themselves, and understand that controlling food so extremely and being so underweight is not what they want or need. But they cannot stop thinking along these lines or restricting their eating.

The usual approach is to help the person increase their weight so they are physically less unwell, using drugs such as the antipsychotic olanzapine and the SSRI fluoxetine (Prozac), which can lead to weight gain as well as improve mood and reduce obsessional thinking.

But too often, treatment doesn't work. Anorexia has the highest mortality rate of all mental disorders.[18]

This makes an experimental treatment worth trying. After our depression studies at Imperial College, we decided to study anorexia, with the hope that psilocybin-assisted therapy would disrupt the disorder's typical rigid, self-destructive mode of thinking. "We need to reframe the question," says Dr. Meg Spriggs, who led the trial. "It's not necessarily, 'How do we treat anorexia?' Instead, the problem we need to address is, 'How do we help people engage in the process of recovery?'"[19]

We were fortunate to get funding from a philanthropist and we are now in the middle of a trial.[20] Several other groups, both in the U.S. and Australia, have started work in anorexia too. The Australian government has made available 15 million Australian dollars (approximately U.S. $10 million) for psychedelic research, including an anorexia study at Sydney University. We've collaborated on the study design, so that eventually we can fuse their data set with ours.[21]

There is also a trial under way at Johns Hopkins that should report in a year or two.[22] And Professor Walter Kaye at the University of California has just reported his study with Compass Pathways; pilot study data showed a reduction in eating disorder and anxiety scores.[23, 24] Based on this and our preliminary findings, Compass Pathways are setting up a multicenter RCT in anorexia nervosa.[25]

It's possible that psychedelics could work in other eating disorders too. Outside the lab there is relevant survey data; one survey of people with eating disorders showed that after taking a psychedelic, people reported an easing of anxiety and depression symptoms and there was also some improvement, although less marked, in the symptoms of the eating disorder.[26]

Another study interviewed people who'd been diagnosed with eating disorders including anorexia and bulimia who had taken ayahuasca. Most reported reduced eating-disorder thoughts and symptoms. It also helped them process painful thoughts and memories and heal the perceived root of the eating disorder.[27]

Tryp Therapeutics is trialing their version of psilocybin, TRP-8802, to treat binge-eating disorder. Preliminary data shows a reduction in daily binge-eating episodes of 80 percent, over a period of four weeks.[28]

There's also some clinical experience that ketamine might be useful in disrupting the fixed thought patterns of anorexia.[29] Awakn clinics have started to offer ketamine treatment for this. One patient reported that under the ketamine trip she was for the first time able to look at herself from outside, and see how distorted her views of her own body had been.

# OBSESSIVE-COMPULSIVE DISORDER (OCD)

Evidence rating: 5/10
OCD was one of the first disorders to be studied in the current era of psychedelic research, back in 2006.

The research took place at a specialist OCD center in the U.S. and was conducted by Professor Francisco Moreno. The study was prompted by anecdotal reports of people who used psychedelics experimentally and found that their OCD symptoms reduced. The researchers theorized that because current treatments such as SSRIs work on the serotonin system, psychedelics that also work on this system might be effective. In a small study, they found that giving psilocybin did lead to a reduction in symptoms that usually lasted more than 24 hours.[30]

Now that we know more about how psychedelics work, it makes sense that they'd work in OCD. It's a disorder where people's actions come from unhelpful and compulsive thinking patterns. Typically, people will go to the doctor for help with a compulsion to perform an action excessively or a certain number of times, such as having to wash their hands or check electrical switches, or check the front door is locked.

The fear underneath the compulsion is that the person believes that if they make a mistake in acting out their compulsion, some form of harm will result. That might be contamination by germs in the case of handwashing or, in the case of electrical switches, causing a fire.

If people can't act out their compulsions, they become very anxious and can get very distressed. This forces their families, who don't like to see their loved one upset, to comply with their behaviors.

Professor Moreno is now resurrecting this line of research at the University of Arizona, in a double-blind trial comparing psilocybin with the sedative lorazepam. Preliminary results show that psilocybin leads to significant improvements in OCD symptoms.[31, 32]

Researchers at Yale University also have an ongoing OCD trial, using psilocybin. Ben, a participant in the trial, says:

The only way to really describe it is, my old habits and behaviors felt vestigial, really like an appendage that didn't make any sense any more. I made this conscious choice to reinhabit my body but my brain was holding on to all these patterns that no longer really made sense to it.

And so as each symptom came up, I took it at face value, realized it no longer applied to me, and they just fell away. One by one, just fell away.

It ended up taking weeks for certain scenarios to occur, symptoms tied to particular objects or that were just less common for me...These things were just like little gifts that appeared over days and weeks. It came to a point where I really felt like after encountering these once, I was completely OCD-free.[33]

A few years ago I was approached at Imperial College by Orchard OCD, a charity that specializes in raising money for OCD research. They suggested we could apply for a grant to replicate and extend the earlier psychedelic work. I was very interested because by then I had seen a remarkable video of a woman who uses mushroom tea to help with her OCD. She says that one trip relieves her symptoms for a few months.[34]

Working in collaboration with Britain's leading OCD expert, Professor Naomi Fineberg at the University of Hertfordshire, we asked patient experts to contribute to the design of the study. They told us they were enthusiastic about the prospect of a potentially powerful novel treatment, but they didn't want to undergo a full psychedelic trip, because they didn't want to lose control.

The compromise we agreed was a midi-dose of psilocybin, 10mg. This isn't enough to produce hallucinations or loss of control but is likely to produce relaxation of the patient's anxiety so they can more effectively engage in the accompanying behavior therapy that's the current therapy for OCD. This kind of midi-dose was used very widely in the 1960s and 1970s, as a prelude to individual psychotherapy sessions, when it was called psycholytic therapy. We have ethical approval and, since we announced it, we have had over 100 patients apply to take part. Our study has now started.[35]

# ATTENTION DEFICIT HYPERACTIVITY DISORDER (ADHD) AND ATTENTION DEFICIT DISORDER (ADD)

Evidence rating: 3/10

It seems counterintuitive that drugs that target conditions where people think too much might be useful for one where people tend to act without thinking. A typical (there is a wide variation) person with ADHD tends to be impulsive, jump around from topic to topic looking for stimulation and interest, do poorly at school, find it hard to organize their life and keep losing things. I had one patient who lost his passport three times during a single trip. The comic Rory Bremner was diagnosed with ADHD as an adult. This is how he describes his life premedication: "It's a bit like plate-spinning sometimes, I feel like my life used to be like how you see circus acts with six plates spinning and there's always one just about to fall off."[36]

To improve attention and concentration, the usual treatment is a stimulant, for example Ritalin (methylphenidate) or another amphetamine. But Albert Hofmann, who discovered LSD, thought that low doses of LSD might be an alternative to Ritalin.[37]

At any one time, ADHD is the most common psychiatric diagnosis in Western countries, around 3 to 4 percent of the population, higher than depression or anxiety.[38] This is because while other conditions come and go, ADHD is permanent.

Not every patient has the hyperactivity—H—part of ADHD; this is particularly true for girls and women, who tend to be more inattentive than hyperactive but do have the churning thoughts we mention below.

There is research suggesting that LSD might have some benefits. Surveys report that some people prefer to self-medicate by microdosing than take the usual medications.[39]

Professor Kim Kuypers at Maastricht University has studied the effects of different doses of LSD on attention in healthy volunteers. She found that lowish non-psychedelic doses of LSD did improve attention.[40]

These psychedelic studies build on the work of Philip Asherson, professor of molecular psychiatry at King's College, London, who pioneered research into treating ADHD symptoms with cannabis, another treatment that seems paradoxical.[41]

Asherson's new approach aimed to treat a rarely discussed but common and debilitating symptom of ADHD—excessive mind wandering and repetitive churning of thoughts.[42]

This is what's probably disrupted by both psychedelics and cannabis. So far, trials of psychedelics have been short-term, so it's not yet clear if they will turn out to be a useful long-term treatment. However, there is commercial interest in continuing this line of research, and a trial of low-dose LSD in adults with ADHD is currently under way.[43]

# Chapter 11

# MICRODOSING, MIDI-DOSING AND MACRO-DOSING

**HOW DO YOU** know if someone is microdosing? They'll probably tell you. People who get into microdosing often become evangelical about it. They might tell you that their energy levels are higher, they can concentrate better, they are more creative and productive. Or they might say microdosing has relieved their stress, lifted their depression, calmed their anxiety or even relieved their migraine.

In the past ten or so years, taking a dose of psychedelics so low that it doesn't produce trippy symptoms, or indeed any obvious effects at all, has rocketed in popularity. It's not surprising: it apparently promises to give you all the above benefits but without the time needed for or the challenge of a full trip. It's had such brilliant PR, it's likely there are now more microdoses being taken than trip doses.

The trend started in the tech community, then went mainstream, spawning thousands of blog and social media posts, forum threads and press articles. One headline written by a mom says: "It makes me enjoy playing with the kids." Another reads: "The TikTokers who are 'microdosing' magic mushrooms and LSD—and say it helps their mental health."[1]

Despite psychedelics being illegal in most places, microdosing is now mostly written and talked about as another wellness hack or lifestyle upgrade, alongside changes in nutrition, meditation and exercise regimes.

You can trace the trend back to 2011, to one man and his book: US psychologist and researcher James Fadiman and *The Psychedelic Explorer's Guide*, which includes a chapter on microdosing.[2]

Fadiman's career spans both the first wave of psychedelics and this one. In 1960, he was turned on to LSD in Paris by his then tutor, psychologist Richard Alpert (later Ram Dass). "I spent the next few hours learning that my worldview had been somewhat limited," he says. He left Paris to meet Aldous Huxley and Timothy Leary, who were speaking in Copenhagen. Back in the U.S., Fadiman took 200mcg of LSD—a very big dose—with two more key figures in the 1960s psychedelic movement, researchers Dr. Willis Harman and Myron Stolaroff. He said it "changed most of my life, the nature of self-identity and time and space." He began working with Stolaroff on psychedelic therapy research at the International Foundation for Advanced Study as a graduate student. The day the LSD ban came into effect, a government letter saying that all studies must stop immediately arrived in the clinic. But Fadiman had just given four scientists a dose of LSD for a study on creativity. The researchers agreed to pretend the letter arrived the following day.

In 2000, Fadiman began connecting with people who, because of the ban, were microdosing, and collecting anecdotal reports. He traces this work back to that first big 200mcg trip, which, he says, "shifted my awareness of the possibilities for human beings. That is the work that I'm still doing."[3]

He set up a website for people to share information on microdosing, describing dosage and effects and advising on safety. It now includes over 2,000 contributors from 59 countries.[4]

At first, the conversation around microdosing was all about improving productivity and creativity. Then, people started to talk about microdosing for mood improvement. Writer Ayelet Waldman's bestseller, *A Really Good Day*, is an account of her 30-day trial of microdosing LSD; it describes how it helped stabilize her hormonally reactive mood during her menopause and, in the process, improved her marriage and her relationship with her children as well as her work output. She's now making it into a TV series.

And yet—you may be surprised to hear—despite how mainstream microdosing has become, despite the thousands of mainly positive personal accounts you can find online, there's no scientific agreement of how it works. Or even if it works.

## SOME REPORTED BENEFITS OF MICRODOSING

These are some of the benefits that people have attributed to microdosing:

* Psychological: positive mood and decreased depression symptoms. Less anxiety. Emotional balance. More energy, empathy. Gaining insight into the self.
* Pain relief: treating cluster headaches and migraine.
* Dependence: decreased craving for substances or use.
* Cognitive: increased work effectiveness, better concentration, focus or mental clarity.
* Creative: coming up with new ideas, finding new thought processes.
* Spiritual: finding meaning.
* Relationships: increased sociability, connectedness and sensitivity to others, better relationships.
* General well-being: improvement in premenstrual symptoms, better sleep quality, healthier eating habits, improved sexual function.[5, 6]

# WHAT IS MICRODOSING?

Microdosing is generally accepted to be taking one-twentieth to one-tenth of a recreational dose on a regular basis, most usually psilocybin or LSD, less often ayahuasca or ibogaine. It's been defined as using subthreshold doses of psychedelic drugs in order to enhance cognition, to boost energy, to promote emotional balance, and to treat anxiety.

In theory, the dosage should be so low that you don't feel any changes in perception or consciousness. In practice, people often use a dose they can detect, perhaps that makes them feel more relaxed or a little more social or in a lighter mood, or even a little light-headed, with vision that's slightly wobbly or sounds that are slightly distorted. This is a dose that they can perceive but that doesn't impact their ability to function.

A typical regime is to microdose two to three times a week, usually at the beginning of a workday, for a block of four to eight weeks, then to take two or more weeks' rest. However, there is a huge variation in regimes, although most people do schedule gaps between days and then gaps between blocks to prevent tolerance.

## HOW DOES A MICRODOSE COMPARE TO OTHER DOSAGES?

The strength of the effects of a psychedelic is determined by the percentage of 2A receptors it occupies in the brain. Thanks to imaging research done at the University of Copenhagen, we have very good data for psilocybin. It shows that a macro-(trip) dose of psilocybin occupies a large percent of these receptors, and lower doses proportionately less.[7]

### Microdose

This dose range leads to slight alterations in perception or consciousness, feeling more relaxed, in a better mood, a little light-headed.

### Midi-dose

This dose range is being used in some studies where people don't want a full trip, for example, in studies on people with early dementia.[8]

In the 1950s, when British psychiatrist Dr. Ronald Sandison was doing his psycholytic psychotherapy to help patients who'd become "stuck" in therapy, as we saw in Chapter 6, he used this dose range too.[9]

At this dosage, stability of vision alters, color sensitivity begins to change, hearing changes and there is more sensation from touch. There are more lucid, wandering, flexible thoughts, perhaps including thoughts of the past. And there may be a sense of dealing with things in more of an emotional way.

In our first depression study at Imperial College, patients with resistant depression were each given two doses of psilocybin: a midi-dose as a safety test and then a macro-dose.[10] On the midi-dose, most people reported feeling different. One had quite a powerful experience of reflecting on her relationship with her sister. Most people will feel more insightful at this dose but won't experience these kinds of breakthroughs.

### Macro-dose

This is a full trip, both as it is used recreationally and the dosage range now most commonly used in research. As described in Chapter 2, perceptions distort enormously, including sounds, touch and visuals. Most people will experience hallucinations that look like Christmas tree lights, sparkling and moving, very clear and bright and appealing. Some people have synesthesia, where the senses hybridize, so that sounds produce colors in the mind and numbers become colorful.

Higher doses than this may not be advisable. For example, research shows that going beyond 125mcg of LSD is unlikely to give more of the desired effects, but it will give more adverse effects, such as headaches, blood pressure changes and anxiety.[11]

## COULD MICRODOSING BE MAGIC?

When you read the extensive list of what people report microdosing has done for them, it becomes clear why it's so popular. But as a scientist, a broad list like this is more likely to make me skeptical. It's hard to research: How do you design a test for so many effects? When I first heard about microdosing, I was very interested in what people were saying, but I found it hard to believe that one substance or even family of substances could do so many things.

Until the past couple of years, there were almost no formal studies (there are still very few). This isn't surprising. In countries and states where psychedelics are illegal—most of them—a single molecule is as illegal as a kilo, so there are the same barriers to research if you want to use only tiny amounts in a study as there are with larger-dose studies: namely stigma and reputational risk, very high costs, and safety- and ethics-related administrative red tape.

At Imperial College, we managed to get ethical permission and some funding from the Beckley Foundation for an LSD microdosing study. But we couldn't afford to do it. The legal requirements and the safety rules imposed by the ethics committee were the same as if we were using a full dose: locking drugs in the safe, giving the drug only in hospital, keeping the participant in for a whole day. The difference is that for microdosing, you have to dose each subject multiple times, and as each would have required a hospital admission this made the study prohibitively expensive. That's why most microdosing studies are survey based, with people using their own drugs in a naturalistic setting. However, there's a fundamental downside to naturalistic studies: it's not possible for the researcher to be certain how much (if any) of the active substance is being used.

## ARE THERE ANY RISKS TO MICRODOSING?

As well as acting on the serotonin 2A receptor, classic psychedelics also act on the serotonin 2B receptor, including in the cardiovascular system. This has been raised as a cause for concern. In the 1990s, a weight loss drug called fen-phen, which also stimulated the 2B, was shown to cause a thickening of the walls of the heart valves and blood vessels that, like pondweed choking a pond, blocked the pulmonary arteries. One-third of patients developed a fatal heart valve abnormality.

In most of the treatments currently being researched, psychedelics are given infrequently, often only once or twice. This kind of irregular dosing regime is extremely unlikely to cause heart damage; it's not frequent enough. And the good news is, despite some people microdosing for some years, nobody has yet reported this as a side effect. Even so, there are companies developing compounds that don't have the 2B effect. In the long term, it's not yet clear whether microdosing will throw up other unexpected adverse effects.

## WHAT MIGHT MICRODOSING DO IN THE BRAIN?

We know from studies of full doses that the main mode of action of the classic psychedelics, including LSD, psilocybin and ayahuasca, is at the serotonin 2A receptor. But there's very limited data on the brain basis of microdosing. It has been suggested that very low doses might improve brain function by increasing resilience to stress, as serotonin seems to do in general. Microdosing might also work on the nerves in the gut or even on the microbiome (gut bugs) to promote well-being.[12]

Research has shown that at full doses, stimulation of 2A leads to neuroplasticity (see Chapter 5), the brain becoming more flexible by neurons growing new connections.[13]

It's not proven if this happens at subtle doses as well, but some researchers have suggested that microdosing might promote the growth of new neurons and so may even protect against brain damage and dementia.

Certainly the longevity of some of the earliest users of LSD suggests that premature death and "fried brains" are not an inevitable consequence of psychedelic use. Albert Hofmann and one of the first medical researchers, the UK LSD pioneer Joel Elkes, professor of experimental psychiatry at the University of Birmingham, both lived to be over 100 years of age. When I remarked on this to one famous but long-retired professor of psychiatry, still sprightly and active in his eighties, he smiled and commented that this must explain why he was doing so well too!

There are 14 types of serotonin receptors besides the 2A, and it's been suggested that effects at some of these other receptors might be key in microdosing. For example, the

5HT6 and 5HT7 receptors are involved in mood and cognition, with the latter also involved in circadian rhythms.[14]

However, Professor Matthias Liechti, the leading human clinical pharmacologist in psychedelics, is skeptical about the beneficial effect of lower doses on, for example, cognition. He has written: "There is no apparent evidence that LSD, which impairs cognition at active doses, would magically enhance concentration when used at lower doses as described by users."[15]

## WHAT DOES THE RESEARCH SAY?

There are thousands of anecdotal reports of microdosing online. But as the adage goes, the plural of anecdote isn't data. That's not to say people's accounts are invalid, but anyone who bothers to write about their experience is more likely to be someone who is more positive about it, and so there is bias built in.

Offering a slightly better quality of evidence, there are observational studies. These are surveys that follow people before, during and after their own microdosing regime, where the data is collected in real time.

For example, a 2019 study followed 98 people starting a microdosing regime and asked them to do daily tests of mood, attention and well-being, as well as asking about aspects of personality, creativity and sense of agency. On microdosing days, people reported better psychological functioning. And at the end of the study they reported less mind wandering and lower levels of depression and stress.[16]

A 2022 survey study had two age and gender-matched groups: one spent around a month microdosing, the other did not. Everyone had a psychological assessment at the beginning, then another one between 22 and 35 days afterward. The assessment asked about addiction, depression and anxiety and tested cognitive skills, memory and processing speed (how fast you can do a task). The people who'd microdosed showed better mood and reduced symptoms of anxiety, depression and stress.[17]

However this kind of evidence isn't considered enough to establish that a drug or intervention works. The scientific gold standard is the double-blind, randomized controlled trial (RCT—see box below). This measures the effects over and above the placebo effect.

## WHAT IS THE PLACEBO EFFECT?

It's when a treatment with zero activity—a sugar pill or a fake injection—leads to someone's symptoms improving. There's a certain amount of placebo effect in all medical treatment; it comes from factors like the person's hope and expectation of it working, their consultation with the doctor and even being in a hospital. The placebo effect can be surprisingly powerful. It can affect someone's experience of taking the drug, their symptoms, their behavior and their physiological response.[18] For example, in some studies the placebo effect has come out better than painkillers including ibuprofen and morphine.[19]

Generally, before a medicine is approved it needs to show in trials that it performs better than a placebo. This is typically done by a double-blind randomized controlled trial (RCT), where some people are randomized without their knowledge to the placebo, the others to the "active" drug group. Neither the researchers nor the subjects know which group they are in (this is what "double-blind" means). This is supposed to stop any knowledge of having been given the drug—the expectation effect—affecting the result.

Recently, imaging studies have shown that placebos produce a real neurological effect, involving several different parts of the brain. "For years, we thought of the placebo effect as the work of imagination," says Kathryn T. Hall, a geneticist and an assistant professor of medicine at Harvard Medical School. "Now through imaging you can literally see the brain lighting up when you give someone a sugar pill."[20, 21]

Hall's theory is that the signals of the placebo effect travel along a chemical pathway in the brain, turning on real biological healing processes, for example, relieving pain. It's been shown that the placebo response is stronger in neurological and psychological conditions, which is exactly in the sweet spot of psychedelics. But I wonder if it may turn out that psychedelics themselves magnify the placebo chemical pathway in the brain.

For a macro-dose of psychedelics, it's impossible to double-blind, as almost everyone will know when they've had the real thing. The therapists or guides can't help but know too. But for microdosing double-blind is theoretically possible, although studies have shown it's not usually perfect. The small but perceptible effects may mean that people are more likely to guess correctly which group they are in. However, there are ways to analyze the data to eliminate this bias.[22]

Recently, there have been some double-blind lab studies in microdosing. The results, for anyone who believes in microdosing, have been disappointing.

For example, one study involved healthy people taking microdoses of LSD four times in the lab, at 3- to 4-day intervals. Each person was randomly assigned to receive either a placebo, a small mini-dose or a larger midi-dose. After taking this, they did an hourly mood questionnaire over five hours, as well as cognitive and behavioral tests.[23]

However, apart from people feeling more positive the first time they had the larger midi-dose the study found no significant effects on mood or cognition. And the larger midi-dose, arguably, is detectable, and so any effects might be down to what the subjects expected.

At Imperial College, we wanted to find out more about the impact of expectations. The team surveyed people before they began their own microdosing regime to ask them what they thought they'd get from it. After four weeks of their regime, they surveyed them again. People reported that microdosing had improved their well-being, mood and performance. They said it reduced symptoms of anxiety and depression and increased resilience, social connectedness, agreeableness and psychological flexibility.

However, there was a kicker: the subjects' positive expectations predicted their positive effects, suggesting a significant placebo response at work.[24]

This is not unexpected: we know that the "set"—mindset—including expectations, is important when it comes to the effects of a full dose of psychedelics.[25]

However, in our second depression study we showed that after two macro-doses of psilocybin, expectation did not predict the outcome. Interestingly, expectation did predict the outcome of taking escitalopram daily for 43 days.[26]

## THE POWER OF PLACEBO

Because we didn't have the funds for a hospital-based microdosing trial, our Imperial College team devised a new kind of naturalistic trial, one that was placebo controlled and randomized too.[27]

Dr. Balázs Szigeti and Dr. David Erritzoe wrote an instruction manual telling microdosers how to create their own placebo control using vitamin capsule shells and envelopes. They were told to put their microdoses into opaque gel capsules but to leave some other capsules empty. These were used as placebos. They then made up eight weeks' worth of capsules—four with the real drug, four with the placebo—and put each into envelopes marked with QR codes.

These QR codes were then scanned, and that connected with the IT infrastructure of the project, telling the participants what to do with the capsules inside that envelope—either to discard it or use it at specific times during the experiment.

It wasn't a perfect trial: we didn't know the dosage as we couldn't test their pills. On specific days, they filled in the questionnaires to rate their mood and experience.

What we found was surprising. After four weeks, as has been shown in numerous studies, we saw that microdosers improved in a number of psychological parameters, such as well-being and mindfulness. However, that was true whether they had been microdosing or taking the placebo; there was no difference between the two groups.

This showed that the expectation of taking a microdose was as good as taking one, a strong placebo effect. It was so strong that a lot of people who'd taken the placebo simply couldn't believe it was the placebo. One emailed: "I have just checked the remaining envelopes and it appears that I was indeed taking placebos throughout the trial. I'm quite astonished [...] It seems I was able to generate a powerful 'altered consciousness' experience based only the expectation around the possibility of a microdose."

It seems microdosing does have a positive effect, but this effect is not larger than the placebo effect. "When we started the study, our vision was that we'd be the heroes to prove that microdosing works better than placebo. Unfortunately our results were disappointing, as we found the opposite," says Szigeti. "The microdosing community was pretty bummed out as well. We even received hate mails when the results were published."

Our conclusion? Microdosing does work...but people's beliefs about it are more important than the drug itself.

That doesn't mean microdosing has no use. A lot of us take vitamins with no evidence that they work, but we think we're benefiting, and so we do benefit.

There is a placebo-controlled lab study that's come out with some positive results. In a lab, 24 people were given either a placebo, a small microdose, a large microdose or a midi-dose of LSD. They were then tested on mood and cognition.

Even with the very small microdose, effects were noted, including improved mood and attention. (The attention effects prompted this group to pursue a study of low-dose LSD in ADHD.)

With the midi-dose there were improvements over and above placebo in mood markers, friendliness and focus. However, people also reported increased anxiety and confusion.[28]

You could even argue that none of the above doses are microdoses, as people were able to detect an effect, even at the lowest dose.

Currently, there's no solid evidence that microdosing works. But neither does it seem to be associated with any worrying negative effects so far, although there have been no long-term laboratory studies. This lack of research means there are lots of other variables that haven't yet been tested in trials. It may be that microdosing works when you take it for longer than has so far been tested in research. Also, dosage is a big unknown in real-world trials, as it's usually not possible to test what people are taking. It could be that microdosing only works for people who have a preexisting condition, such as depression; if these people have low serotonin systems, a microdose may be enough to boost it.[29]

An Australian company, Woke Pharmaceuticals, is starting the first trial to test regular microdoses of psilocybin to treat depression, recruiting over 250 people with moderate depression. Half will microdose for six weeks, and half will be given a placebo. Dosage will be on the high side for a microdose.[30]

## WHAT HAPPENS WHEN CHILDREN TAKE LOW DOSES?

In the Brazilian religions Santo Daime and União do Vegetal (which we met in Chapter 2), the service includes dancing, chanting, and taking a weekly, lowish dose of ayahuasca. The function of the ritual is to bring the community together. In these churches the ritual use of ayahuasca often begins in late childhood, although it can be earlier. Research shows that the children who take ayahuasca aren't negatively affected. Indeed, the reverse might be true; a comparison of around 40 teenagers from one of these groups to 40 control teens showed that there was no difference in cognitive abilities, including verbal and visual abilities, mental flexibility and memory.[31] And compared to controls the teens who took ayahuasca scored lower on anxiety, body dysmorphia and attentional problems.[32] The ayahuasca-taking teens also used less alcohol and were less likely to have used it recently.[33]

This brings up an interesting question. Adolescents don't always do well on the usual antidepressants. Might low doses of psychedelics turn out to be beneficial for treating adolescents with mental illness, or even help them learn to deal with stress or conflict? This is something we are beginning to research at Imperial College.

## Chapter 12

# MY SACKING, MDMA AND THE REAL HARM OF DRUGS

**IF DRUG LAWS** are to do their job—that is, deter people from using these substances—they need to be based on real evidence. In other words, the punishment for using a drug needs to be proportionate to its harm. Now that MDMA has been deemed safe enough to be a medicine (in Australia) and will likely be soon in the U.S. and Canada as well, it has become even more clear that the way ecstasy was portrayed by the UK government and the media throughout the 1990s and 2000s was not based on science.

In the 1990s, I began doing work for the UK government, joining the Advisory Council on the Misuse of Drugs (ACMD). The 1971 Misuse of Drugs Act classifies drugs in three levels—A, B or C—based on their relative harm, and these are supposed to be decided by the ACMD. The Act was designed to remove decision-making about drugs from party politics, to minimize the risk that short-term party interests might lead to bad laws. MDMA was in Class A alongside heroin and cocaine, the class for the most dangerous drugs with the highest penalty.

At the time the press was very antidrugs, in particular anti-ecstasy. Articles said that ecstasy caused lasting brain damage, memory damage and depression. The press reported every death where the person had taken the drug, often in lurid detail. Despite these sad deaths, a lot of experts from all kinds of backgrounds—the police, medical, science, some politicians and me—began to reach a consensus that MDMA simply wasn't addictive or harmful enough to be in Class A.

But it became clear that the government was not committed to a wholly evidence-based drugs policy, in particular when it came to MDMA and cannabis (for the story of the latter, see my book *Cannabis*). The UK government kept to a War on Drugs agenda: all illegal drugs are dangerous, prohibition will stop drug use, drug users are criminals.

I repeatedly witnessed politicians making decisions about drugs for political reasons. For example, in July 2005, as a knee-jerk reaction to press reports of head shops selling freeze-dried magic mushrooms, the government, led by Tony Blair, rushed through legislation that put them into Class A.

This was illogical; millions of people have taken mushrooms for thousands of years, with very few reports of negative consequences. But the government wanted to score points by being seen as "tough on drugs" as the opposition Tory party said they would be. And they didn't even give the ACMD an opportunity to do what we were set up to do: consider the evidence.

In the 2006 report *Drug Classification: Making a Hash of It?*,[1] the House of Commons Science and Technology Select

Committee suggested the ACMD do a full review of the classification of MDMA, with a view to possibly moving it to Class B. One reason given was this: "Recent figures show that there were about 13.5 times more ecstasy users than heroin users in 2004, and deaths caused by ecstasy were around 3% of the number caused by heroin."

Over the course of several ACMD meetings, we created the first evidence-based assessment of the harm of the most commonly used drugs, including MDMA. Taking each drug in turn, we used the same methodology to score it on its harm. We published this analysis as a paper, "Development of a Rational Scale to Assess the Harm of Drugs of Potential Misuse," in *The Lancet*. It clearly showed that Class A didn't reflect ecstasy's harm. In fact, it showed that the only drugs less harmful than ecstasy were alkyl nitrites (amyl nitrite or poppers) and khat.[2]

When the ACMD published our 2008 report on ecstasy[3] we made a number of recommendations, most of which were accepted by the government. But two that were key were rejected. One was to establish a national testing scheme for harm reduction and trend monitoring purposes. The second was to reclassify MDMA from Class A to Class B in recognition of its lower relative-harm profile. The home secretary Jacqui Smith refused to countenance that too, despite all the studies and statistics we'd examined, and the opinions we'd gathered from a wide range of experts.

I began to be more outspoken in my criticism of the government's drug policy. By this time, I had spent ten years working on the ACMD, a committee supposedly developing the evidential base for drug laws. But all that work was going to be ignored.

Think how this felt from the perspective of a scientist. Science depends on the fundamental principle of relying on evidence (aka the truth) to make decisions. And the government were getting away with policies to appease the right-wing media that were actually increasing harm. One example of this is that when substances are banned, individuals get around the law by making new, often more potent and more dangerous legal alternatives.

I wrote a thought piece about a made-up addiction called Equasy,[4] short for equine addiction syndrome. One of my patients had suffered irreversible brain damage from falling off her horse, which had got me wondering how dangerous riding might be. I discovered that it was more dangerous than I had thought, particularly over jumps.

The point of the article was not that people should take ecstasy rather than ride. Or that they shouldn't ride. Or that ecstasy wasn't harmful. I wanted to educate people about drug harm in relation to the harm of other activities in life, making it possible for them to make sensible decisions about relative harm.[5]

This was not acceptable to the government. The home secretary telephoned me, shouting that I had exceeded my position as ACMD chair by comparing a legal activity with an illegal one. I tried to explain that I thought riding was a good comparison as another thing young people do that is also dangerous. Especially as the public tend have the perception, driven by the media, that ecstasy is more dangerous than it actually is. She didn't listen. Perhaps she thought that publicly criticizing me might deflect attention from her abuse of parliamentary expenses.[6,7]

I gave a lecture at the Centre for Crime and Justice Studies, called Estimating Drug Harms: A Risky Business. I called for a new way of classifying the harm caused by both legal and illegal drugs, explaining that, in our *Lancet* study, alcohol ranked as the fifth most harmful drug after heroin, cocaine, barbiturates and methadone. Tobacco came in at ninth most harmful. Cannabis, LSD and ecstasy, while harmful, ranked very much lower at 11, 14 and 18 respectively.

I said that we had to accept that scaring young people didn't stop them from experimenting with drugs and other harmful activities. And that the government's job was to protect them from harm. And that the only information that they'd listen to was credible, accurate information. I said: "We have to tell them the truth...A fully scientifically based Misuse of Drugs Act where drug classification accurately reflects harms would be a powerful educational tool."[8]

The new home secretary, Alan Johnson, didn't agree. He saw my speech as briefing against the government; he asked for my resignation, then sacked me. Johnson wrote the following in a letter to the *Guardian*: "He was asked to go because he cannot be both a government adviser and a campaigner against government policy. This principle is well understood and long established."[9]

## REAL DANGERS MATTER

After I was sacked, I continued my work measuring the real harm of drugs along with a group of eminent experts and started my charity, Drug Science. In 2010, we published a more sophisticated analysis of drug harm, using a technique called Multi-Criteria Decision Analysis (MCDA), designed for very complex areas. We plugged in expert opinions on the whole spectrum of 16 harms of each drug, both to the individual and to others, including dependence, harm to health, crime and economic and environmental damage, and MCDA turned them into a definitive harm score for each drug.[10]

As you can see in Figure 13, when harm to both the user and society were included, alcohol is the most harmful drug overall. If you look at just harm to the user (the top bar),

## Figure 13: Drugs ordered by their overall harm score

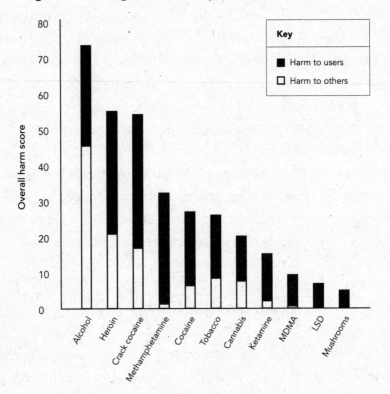

Source: Adapted from Nutt DJ, King LA, Phillips LD, 2010, Drug Harms in the UK: A Multicriteria Decision Analysis, *LANCET* 376: 1558–1565, ISSN: 0140-6736.

crack comes out most harmful, followed by methampheta-mine and heroin.

The psychedelic and psychedelic-like drugs we looked at— ketamine, mushrooms, LSD and ecstasy—came out as very low in harm both to others and to the user. In particular, LSD and mushrooms were extremely low on both.

The rest of this chapter digs down into the risks of MDMA and other psychedelics (excluding dependence, which is in Chapter 13). Bear in mind that most of the research on harm is on drugs used in an illicit context and that the way a substance is used in a controlled trial or medical setting is always exponentially safer than illicit drugs that are taken at a club or a festival. In trials, we use a guaranteed dose of a pure compound, we make sure the patient or participant is suita-ble, we prepare them as to what to expect, we give the drug with at least one and usually two therapists present, and we help the patient integrate their experience afterward.

## MORE WAYS TO LOOK AT DRUG DANGERS

Another way of looking at how dangerous a drug is is using a measure called the margin of exposure, a ratio between the level that causes an overdose and the amount people usually consume or take. It's used to calculate, for example, the risk of amounts of carcinogens in food. For cannabis, it would take hundreds of times the average dose, i.e., there is no possibility of overdose. Twice the average dose of heroin can kill you. MDMA is around ten times, in compari-son to cocaine, which is about two times.[11]

There's another measure of danger: the index of fatal toxicity. It's the absolute number of fatal poisonings caused by a particular drug divided by its consumption figure. It's often used to show the relative risks of prescription medicines.

On this scale, heroin comes out highest, around 30 times more dangerous than ketamine and MDMA.[12]

## WHY DO PEOPLE DIE OF MDMA?

In the UK, probably the first MDMA-related death that everyone remembers was that of Leah Betts in 1995. She died in the hospital five days after her eighteenth birthday. She'd taken one ecstasy tablet at home with some friends. Her parents released a photograph of Leah on life support in a hospital bed, taken just before she died. Her image was used in a poster campaign and a school drug education program. A generation of teenagers heard the slogan, "One pill can kill."

Dr. John Henry, a toxicologist at a west London hospital emergency department and a pioneer in MDMA research, gave evidence at Leah Betts's inquest. In the early 1990s, Henry had treated MDMA-related hospital admissions and published papers on why otherwise healthy young people were dying from this new dance drug. He'd identified some key MDMA related complications: overheating, dehydration and delayed kidney functioning.

At the time of Leah's death, clubbers were being warned not to get too hot, to take breaks and to stay hydrated. At the inquest, Henry's evidence said that Leah didn't die from taking an adulterated tablet or having a toxic reaction to the drug itself but from drinking too much water—hyponatremia —rather than drinking too little. She drank 7 liters of water in under 90 minutes in a non-club setting, a tragic illustration of following harm-reduction advice leading to the ultimate harm. After Leah's death, harm reduction advice for people using MDMA changed to include the risk of drinking too much as well as not drinking enough.

## DOSAGE AND PURITY

Initially, MDMA was most usually sold as white tablets. Then, in the mid-2000s, MDMA in crystalline form came onto the market, a higher-priced, higher-purity "premium" product.[13]

A crackdown in Thailand on the production of a key ingredient, safrole, a natural product, led to a global drought in 2008–2010. Of course, drug manufacturers didn't give up and soon developed a new and cheaper synthetic manufacturing process that didn't require safrole. The new tablets were shaped, colored and branded, and they had a boosted MDMA content. Some contained up to 477mg of MDMA, four adult doses. As a result, the MDMA-related death rate also increased in the UK, from 8 deaths in 2010 to 92 deaths in 2018. (It has now gone down but is still running pretty high: 78 in 2019, 67 in 2021.)[14]

A lot of these deaths were of teenagers with relatively little experience, or tolerance, of MDMA and who seem to be more susceptible to the water-retaining effects of MDMA, and so more likely to get water intoxication.

## DOES MDMA DAMAGE THE BRAIN?

In the U.S., MDMA was banned in 1985. After that, the only funding made available for studies on it was to look at its harm. Their focus was usually on the brain. If evidence could show there was brain damage in humans from taking recreational doses, this would justify the ban and scare people off using it.

MDMA releases masses of serotonin in one big rush, which is part of the reason it makes you feel so good. In the normal brain, neural impulses stimulate the release of serotonin, which crosses the synapse to the next neuron's receptor. The serotonin is then sucked back up by transporters so it can be reused. But taking MDMA not only leads to a bigger release of serotonin but less reuptake by transporters too.

The anti-MDMA studies focused on two areas: one, that it caused cognitive damage, such as impairment to executive functioning, the ability to learn and remember. And the second focus was on MDMA leading to a damaged serotonin system, in turn causing depression.

After its prohibition in the U.S., the number of people using MDMA grew. In July 2001, New York police confiscated one million ecstasy pills, then the single largest ecstasy seizure in history.[15] That year, the U.S. Sentencing Commission introduced a new and extreme sentencing guideline for ecstasy too.

In the U.S., sentencing for offenses involving illicit drugs is guided by "marijuana equivalency." Previously, the sentence for ecstasy was, per one gram, the same sentence as for 35g of marijuana. In 2001, the penalty for ecstasy was raised 14-fold. The new penalty was that one gram of ecstasy should get the same sentence as 500g marijuana (one gram of cocaine was equivalent to 200g marijuana and heroin was equivalent to 1000g). One of the key reasons for this huge increase in penalty was the superposition that MDMA is neurotoxic.[16]

One of the key authors of scientific evidence against MDMA was Professor George Ricaurte of Johns Hopkins School of Medicine. He was an established neuropathologist who, with his wife, Professor Una McCann, a psychiatrist, developed a research interest in the possible harm of MDMA on the brain, funded by the U.S. government.

A widely cited 1998 human study by McCann[17] used PET (positron emission tomography) imaging to look at proteins that transport serotonin in the brain in 14 MDMA users, comparing them with 15 non-using controls. They reported that MDMA users had decreased binding of transporters. The subjects had used MDMA between 70 and 400 times.

However, only one of the MDMA users had binding data that was outside the control group range—and this person had taken over 150 doses. And the person with the lowest binding in the control group had lower binding than all but one in the MDMA group.

This suggests that even if MDMA use had reduced binding, the reduction didn't take MDMA users outside the normal range. Also, there was no correlation when it came to dosage —and some subjects reported taking an unimaginably high 1250mg. The authors' claim that MDMA damaged the serotonin system in humans was not proven.

Ricaurte's 2002 study turned to preclinical animal models to see if MDMA could damage the brain. The study was intended to show what happens to the brain after a night out on ecstasy. Monkeys were given the drug, along with loud music (anecdotally, I heard it was the Pogues) and bright lights.

Results suggested that even one night's recreational use of MDMA could result in damage to the dopamine system and thus increase the risk of Parkinson's disease.[18]

When the results were published, Alan Leshner, a highly respected former director of the U.S. National Institute on Drug Abuse, said: "Using Ecstasy is like playing Russian roulette with your brain function."[19]

The trouble was, the *Science* paper just didn't make sense. Millions of people had been using MDMA for years, some even taking large doses in loud music venues, but no cases of Parkinson's disease were being seen. The data was questioned and, months later, the paper was retracted.

It emerged that, due to the incorrect labeling of a bottle, the drug that was actually given wasn't MDMA but methamphetamine, already well known to have toxic effects on the dopamine system in all animal species, including humans.[20]

## RESTORING SANITY

In 2010, the American Civil Liberties Union (ACLU) challenged the heavy sentencing being handed out for MDMA at a two-day *ACLU v. United States* hearing. The original court case was a man, Sean McCarthy, who'd been caught with 70 tablets of MDMA and given a prison sentence of five to six years.

The heart of the argument in court was whether MDMA is (i) neurotoxic, (ii) a hallucinogen and (iii) addictive, the key reasons for adopting the sentencing guidelines.

Professor Val Curran, founder and director of UCL's Clinical Psychopharmacology Unit, argued for the ACLU. She said that although MDMA did have stimulant properties, it was not a hallucinogen, that it had little potential for addiction, and that the evidence for it being neurotoxic was both flawed and inconclusive.

Curran said animal studies of MDMA-induced neurotoxicity do not reflect typical human patterns of ecstasy use.[21] She explained that early animal studies injected 5mg/kg for four days into squirrel monkeys and similar doses into rats. In an average 75kg human that would be a massive 375mg per day, much higher than the oral doses of 80 to 120mg that most people took just one or two days in a month. (And also much higher than the 80 to 120mg people take two to three times during MDMA-assisted therapy.)

She also argued that in human studies, it's hard to tell which effects are caused by MDMA. Studies don't take into account that before taking MDMA people might have differed in their serotonin markers, cognitive function, impulsivity and mental health. And illicit pills can contain contaminants, including other psychoactive drugs. In fact, people who take MDMA often mix it with alcohol and/or cannabis and other drugs. Further, MDMA users at parties or raves use it in the evening and then stay up much of the night because of the drug's stimulant effects. So the following day, they don't function at their best on measures of cognitive performance, partly due to sleep deprivation. And finally, pills vary wildly in the amount of MDMA they contain, and when people report what dose they've used, it can be inaccurate.

For studies that look at brain effects in people, she said the levels of any changes are so small as not to make a difference to people's functioning. And that it looks as if, when people stop using, any effects get better over time.

Curran's evidence was compelling. The trial judge, William H. Pauley III, ruled that the sentence ratio should be lowered by 60 percent so that each gram of MDMA would now be treated as equivalent to 200g of marijuana, not 500g (which made it the same as cocaine).[22, 23]

Judge Pauley concluded that the 2001 hike in sentencing for MDMA reflected a distorted and selective use of scientific evidence. Indeed, he accused the 2001 U.S. Sentencing Commission of "opportunistic rummaging" of facts and concluded that the Commission's analysis was "selective and incomplete," noting that, unlike cocaine, ecstasy is "one of the least addictive drugs."[24]

## DOES MDMA MAKE YOU DEPRESSED?

After the brain releases serotonin on MDMA, it can't replenish its stores instantly; this can take a few days.[25] The result, the so-called Tuesday blues, is probably a combination of the impact of the drug, a night's dancing, lack of sleep, lack of food and dehydration.

The bigger question is, does taking MDMA repeatedly lead to lasting low mood or depression? Ecstasy users do report more depression symptoms. There is evidence that low mood states, depression, irritability and anxiety can go on for months.[26]

Several studies have used PET scanning with different tracers to look for possible changes in the serotonin system as a result of taking MDMA. Some have found a reduction in serotonin transporters and others an increase or a decrease in some serotonin receptors. Both findings could reflect brain compensation for a reduction in serotonin levels. In people who use MDMA, the effects seem to be dose related; more use was associated with lower levels of the transporters.[27, 28, 29]

However, this kind of serotonin brain imaging study has tended to measure the brains of people who use more ecstasy than the average. Professor Adam Winstock cross-referenced the amounts taken by study participants with the amounts people report taking in the Global Drugs Survey. He said the amount the study participants used put them in the top 5 to 10 percent of users in terms both of how much they took and how often. In fact they consumed, on average, seven times more pills over a year than the Global Drug Survey participants.[30]

More recent studies in preclinical models, using MDMA in doses equivalent to those used by humans, have not found evidence of significant neurotoxicity.[31] One reason may be that, unlike methamphetamine and other stimulants, MDMA releases more serotonin than dopamine, and this seems to be protective of the brain and also of the heart.

A meta-analysis of brain imaging studies in MDMA users suggests that transporter levels return to levels similar to those of controls over time.[32]

It's still unclear whether there are any lasting depression effects. But there are unlikely to be any when MDMA is used as a medicine. In the BIMA study (see Chapter 7), we looked at this in detail and didn't find any evidence of lasting low mood in subjects after giving them MDMA. And these were a group of people with alcohol dependence, and so potentially vulnerable.[33]

Another way to explore possible effects of MDMA on the brain is using sleep quality. This was the first study that Professor Robin Carhart-Harris did with me at Bristol, for his PhD.

Serotonin pushes back the dreaming phase and so, when you deplete it, dreaming happens earlier in the night. This is one of the most sensitive measures of serotonin function in the brain because even a small drop in serotonin shortens the time to the dreaming phase.

For the study, we compared users (people who'd taken more than 150 ecstasy pills) and non-users (people who'd never taken any). The baseline measurement was their normal sleep. We found that there was no difference in the dream phase between users and non-users, which suggests that users were not serotonin-deficient.

Then we depleted brain serotonin with the tryptophan depletion test. This is the standard technique, involving lowering levels of the amino acid tryptophan via dietary methods. This reduces serotonin production in the brain, and so makes dreaming happen earlier in the night.

We reasoned that if MDMA had damaged the serotonin system in the users, the impact of depleting it would be magnified. But it wasn't. Tryptophan depletion had the same effect in both groups. Overall the conclusion of the study was that MDMA does not appear to alter serotonin functionality in the brain.[34]

When we tried to get this paper published in a U.S. journal, the editor told me to forget it. He said no American reviewer of our paper would accept that MDMA didn't harm the brain! That is a good example of the attitude to MDMA in 2009.

## DOES MDMA RUIN YOUR COGNITIVE FUNCTION?

Most of the cognitive effects of MDMA that were alleged in the 1990s have either not stood up to later studies, or still not been proven and/or repeated.

A 2019 review used the powerful Bradford Hill criteria, widely used in public health research, to evaluate the evidence from epidemiology studies to determine whether MDMA caused all the neurotoxicity it has been accused of. It made the same points as Professor Val Curran had, plus more: that studies didn't properly separate light and heavy users, and that "most meta-analyses have failed to find clinically relevant differences between ecstasy users and controls." And finally, that "there is also consistent evidence of publication bias in this field of research."[35]

If there is one remaining effect that may have some evidence, it is that of MDMA on memory. But even here the effect is extremely small. And it may be that any changes are due to recreational drug use in general, not just MDMA.

A 2007 study looked at memory and learning in four groups: current ecstasy users, ex-ecstasy users (abstinent for at least a year), poly-drug users and never-users. Results showed that in the current users—both ecstasy only and poly-drug—there were some impairments in verbal learning and memory, but none in ex-users.[36]

Another study in the Netherlands did behavioral testing and neuroimaging on a group of young people who were likely to be future MDMA users. Once some of them had tried ecstasy, the study compared a group of around 60 to a similar group of non-users. Most hadn't used much—a mean average of 3.2 tablets and a median of 1.5. On a memory task, MDMA users scored 29.95 out of 30 before using, and 29.66 at follow-up. The controls scored 29.88, then 29.93 in the second test.

The scores were incredibly close. MDMA users remembered only half a word less (from a list of 15) than non-using controls, who were remembering nearly all 15. But in order to find a significant difference, the study authors used a new way of measuring it. They stated, "Although the performance of the group of incident ecstasy users is still within the normal range and the immediate clinical relevance of the observed deficits is limited, long-term negative consequences cannot be excluded."[37]

Now there is much less concern about the possible neurotoxicity of MDMA. It looks as if, when people stop using, any effects on the brain or memory get better over time. We haven't seen large numbers of people developing disorders that might have resulted from serotonin depletion, such as depression. In fact there is growing evidence that one of the ways MDMA may help people with PTSD is by healing the brain through the process of neuroplasticity (see Chapter 5).

## IS MDMA A SAFE MEDICINE?

In order to become a medicine, MDMA has had to go through a long series of trials, which have been extensively scrutinized at every stage. MDMA medicine is very different from street ecstasy, where, most of the time, you can't guarantee what you are taking.

Taken as a whole, the evidence suggests that MDMA may not be suitable as a medicine that's taken regularly, for example, daily, as SSRIs are. However, when given a limited number of times as part of therapy, where the dose is carefully titrated and people aren't taking other drugs, the evidence shows MDMA is safe.

# WHAT ARE THE RISKS OF TAKING ECSTASY?

One in a thousand people have adverse effects from taking MDMA in the form of ecstasy.[38] Your risk goes up the more you take, and if you mix drugs. In Australia, from 2000 to 2018 there were 329 MDMA-related deaths. Half were due to taking multiple drugs, 38 percent to other causes (mainly road accidents) and only 14 percent to MDMA toxicity. People who died due to MDMA toxicity alone had a much higher blood level of MDMA than those who'd taken other drugs.[39]

Also in Australia, 0.6 percent of MDMA users end up in hospital emergency room. "Many of those hospitalizations could have been avoided if the user knew about staying hydrated, safe quantities and not mixing drugs," says Winstock.[40]

## HIGHER DOSAGES

When ecstasy first came out, the average dose was 40mg per tablet. Then dosages began to rise. In 2009, only 3 percent of pills contained over 140mg of MDMA. In 2015, it was 53 percent. In 2018, 72 percent contained over 150mg. Some "superpills" contain 270 to 340mg, up to four times a normal adult dose.[41]

In 2022, the average pill strength was 167mg, but there were some with as much as 280mg.[42]

In 2016, Winstock said: "If you take that dose all at once, most people will find it really unpleasant. They'll come up too strongly, they'll vomit, become confused and hallucinate."

There's also been a rise in people taking MDMA in powder or crystal form, buying it in gram bags. This makes it very hard to judge how much you are taking. And taking a whole gram is definitely an overdose.

Teenagers, particularly young women, who are taking drugs for the first time are most vulnerable. "Most MDMA deaths aren't related to dose, except when it comes to first-time users, in particular teenagers. There's something about adolescence and high doses that's intrinsically dangerous. Young people haven't learned how to take drugs yet. It's easier to overheat, to dehydrate, and maybe take other things. High dosage and inexperience is a really bad cocktail."[43]

## MIXING DRUGS

A study has shown that only one in seven deaths linked to MDMA are due to taking MDMA on its own.[44]

## ADULTERATED DRUGS

During the 2021 festival season, the Loop festival testing service saw a large rise in "imposter" drugs being sold as ecstasy, notably caffeine and stimulants called cathinones, some of which cause insomnia and anxiety. The 2022 season saw a return to pre-pandemic levels of adulteration; 11 percent of MDMA products contained no MDMA.[45]

## HOW DRUG TESTING SAVES LIVES

Two of the main issues that kill people who take recreational drugs—accidental poisoning and overdose—can both be prevented by drug testing services. Testing is effective not only for the person who's having the drug tested, but at detecting any adulterated substances on the market and variations in strength. This information can then be shared.

Testing has been around since the 1960s in the Netherlands but in recent years it's spread to Australasia, as well as North and South America. There are now over 30 NGOs providing this service globally. In New Zealand recently the government created a licensing scheme for organizations doing this work, an idea that could be rolled out globally.

In the UK, the Loop is the main charity. However, often the festival organizer won't use them due to uncertainty about the legal status of doing so and having to get permission from the police and the landowner. They tend to err on the side of not running the risk of breaking the law rather than reducing harm to their customers.

The Loop offers a free, anonymous and nonjudgmental service. Staff include 30 chemists in a mobile laboratory and they can give results about an hour later. Teams include doctors, nurses, pharmacists, psychiatrists, social workers and substance misuse practitioners, so they can provide a wide range of advice.

Freddie Fellowes, founder and owner of the festival Secret Garden Party, says that the Loop "represented the only positive advance in harm reduction I saw in the whole 15 years of running the Secret Garden Party. It cut hospital admissions and prevented potentially fatal misadventure. The fact it isn't, now, a legal requirement of all licensed music events of this nature is a question that should be answered by all in authority."[46]

A study of the Loop's testing services at three English music festivals showed that, in 2017, more than half the people who discovered they'd bought substances that weren't what they expected disposed of them.[47]

## HOW DOES MDMA BECOME FATAL?

- **Increases in heart rate and blood pressure**
  MDMA increases all these, making it dangerous for someone who has a heart condition. That's why people with high blood pressure and heart conditions can't take part in trials using MDMA.
- **Hyperthermia or overheating**
  MDMA upsets the hypothalamus, the brain's temperature regulator, putting you at risk of overheating. This is made worse by long periods of dancing in hot conditions. It's also dose related: the more you take, the more likely this is.
- **Hyponatraemia, aka low sodium in the blood**
  This can lead to seizures and coma. This is often due to

drinking too much water, as MDMA releases a hormone that makes your kidneys work slower.

- **Serotonin syndrome**
A big spike in release of serotonin can directly result in seizures, coma, cardiac arrest and intracranial hemorrhage.
- **An autoimmune reaction**
There have been rare cases of autoimmune hepatitis reported that have led to liver failure. This can happen with any drug and/or medicine. We don't know why it happens with MDMA.[48, 49]

## SENSIBLE GUIDELINES FOR USE OF MDMA, KETAMINE AND PSYCHEDELICS

If you are a parent, and your child is likely to take drugs or has an opportunity to take drugs, make sure they know how to do it safely. Deaths of young people are often because they are ignorant.

Some of these points will be more relevant to some of these drugs than others.

* If you can, get your drugs tested. Ideally, don't consume any pill or powder without testing it.
* If you can't test, try having a very small amount and wait for a few hours to see how you feel.
* Don't mix one drug with any others.
* Don't take when drunk. Most ketamine deaths are from taking it when users were drunk. And don't take sedatives or opiates to "come down" afterward.

* Avoid frequent and heavy use of any drugs.
* Measure your dose. Take care with powder and crystal form; it is very easy to overdose. You need to know the weight. Never take the full packet. Remember, pills can vary hugely in strength.
* Use any drug when you're in a good mood, not to treat low mood.
* Have a friend who is not using who you can go to if you don't feel good.
* Have breaks from dancing to keep your body temperature down.
* Make sure you keep peeing.
* Make sure you have a safe way to get home. Carry emergency money in case you get stuck. Don't get into a car with a drunk or drugged driver.
* With ketamine, be conscious of safety while taking a bath or going outside in the cold. Do not take psychedelics when you're near any heights, for example, cliffs or at the top of a high building.
* Avoid taking drugs if you have any heart condition, blood pressure problems, epilepsy or asthma.
* This advice isn't about psychedelics, but it's important; do not take opiates. And never use a needle to take any drugs.

From the evidence, you can see that drugs do have harms, but also that there are ways to mitigate them, particularly when they're used as medicines and so use and dosage is carefully regulated.

# Chapter 13

# ARE PSYCHEDELICS ADDICTIVE OR DANGEROUS?

**I WON'T FORGET** the first talk I gave at a meeting of experts in addiction on the possibility of using psychedelics to treat addiction. During questions at the end, one of the treatment providers got on the microphone and asked me how I could talk about giving illegal drugs to people who had problems with drug dependency. He told me that it was both outrageous and despicable of me even to suggest it. I was slightly surprised at his vehemence. Afterward, another expert commented to me that, if what studies were showing about the treatment potential were true (see Chapter 7), it would very much undermine established treatment models—and profit margins.

When I speak at events outside the psychedelic research community, I'm often asked a question along the same lines as that original commentator, although thankfully in a less combative way. And it is this: How can it be safe to use a dangerous drug as a medicine? I've been asked this question by addiction specialists and also doctors and psychiatrists; politicians, psychopharmacologists and the general public.

A large proportion of people still think that because psychedelics and psychedelic-like drugs are controlled, they must be dangerous. The reasoning that lies behind this is circular: they think if a drug is controlled it must be dangerous, otherwise it wouldn't have been controlled.

But in the 1960s, psychedelics were banned, not because they were dangerous, but for political reasons. When LSD left the lab and therapy rooms, the youth movement it fueled went against the "square" status quo and changed how young people thought, in particular about being drafted for the Vietnam War.

As the ethnobotanist and philosopher Terence McKenna once said, psychedelics are "catalysts of intellectual dissent." LSD was banned because it changed people's minds, not because it harmed them. McKenna also said: "Psychedelics are illegal not because a loving government is concerned that you may jump out of a third-story window. Psychedelics are illegal because they dissolve opinion structures and culturally laid down models of behavior and information processing. They open you up to the possibility that everything you know is wrong."[1]

After the ban, the U.S. government needed to justify it. And so it promoted questionable science magnifying some of LSD's real risks and fabricating others, and made propaganda aimed at the parents of middle-class, white teenagers and young adults.[2]

Films warned of the dangers of "bad acid" that would see their children sectioned forever, and of LSD causing cancer and birth defects. Posters read: "Will they turn you on...or will they turn on you?"

The banning of LSD led to the banning of similar drugs, such as psilocybin. Then, because of the U.S.'s international influence via the UN, prohibition spread across the world, along with misinformation and myths about these drugs.

Fast-forward to the present day and both magic mushrooms and LSD have been tried and tested on a global scale for 60 years and have caused relatively little harm, considering the number of people who have used them. The same is true of MDMA, which has been used widely for 40 years.

Now that psilocybin and MDMA are medicines, we can dispel the myth that these drugs are "dangerous." They have risks and side effects, as do all medicines, and it's true that people have died or suffered after taking these substances; as you'll see, most of the risk of taking them comes from them being illicit and how and where illicit drugs are taken. The job now is to find out how to use them as medicines in the safest and most effective way possible. That's what the newest wave of studies is achieving, but we can also learn from their long history of use.

## CAN YOU BECOME ADDICTED TO PSYCHEDELICS?

One of the most enduring beliefs about classic psychedelics that I come across is that they are addictive. Even the UN still states that they lead to psychological dependence.[3]

And in the bible of psychiatric diagnosis, the *Diagnostic and Statistical Manual of Mental Disorders (DSM-5)*, dependence on these drugs has its own classification: Other Hallucinogen (LSD, MDMA) Use Disorder.

One definition of addiction is: a behavioral disorder underpinned by changes in the brain, which leads to continued use of a drug or substance in the face of problems such as withdrawal, and your use of that substance interferes with your family and social life and causes you personal harm.

You spend more and more of your time preoccupied by the substance, trying to buy it or using it; you spend more of your money on it. You start to give up other activities in favor of taking it. You may lose your job, or your family relationships may be affected by it.

Typically, dependence includes both tolerance—the effects diminish over time—and withdrawal—there are unpleasant effects when you stop. The classic psychedelics don't fit this definition.[4] Treatment facilities aren't full of people who are dependent on LSD or psilocybin. The key to this is their effect on the brain. Psychedelics do produce a very rapid desensitization or tolerance response. Keep taking LSD and the euphoric and psychedelic effects quickly decrease. However, unlike addictive drugs, taking more does not bring the effects back, i.e., even if you wanted more of the effects (and not everyone does), you couldn't get them.

A study done on the U.S. military found that on the third day of taking LSD, the effects wore off.[5] At the time of this study, the Cold War was at its peak. There was a persistent rumor that LSD was so potent that the enemy could put a single bottle in, for example, the New York water supply and disable its infrastructure. But tolerance kicks in so fast that this isn't possible:[6] after a time, people would simply stop feeling the effects.

The classic test to establish if a substance is addictive is done on animals. If it is addictive, the animals will keep self-administering, but this doesn't happen with psychedelics.[7] And they have no withdrawal symptoms, with one exception: ayahuasca can give some minimal withdrawal symptoms.[8]

In fact, one study concluded that psilocybin has a lower risk of dependence than caffeine.[9]

As you read in Chapter 3, psychedelics are, in fact, anti-addictive; the main reason people use ibogaine is actually to treat addiction, most often to opiates. And psilocybin and LSD are now being trialed to treat addiction to a variety of drugs including tobacco, heroin, cocaine and alcohol.

A patient in a 1973 US study where LSD was used to treat heroin addiction, Leonard N., said: "The two experiences of heroin and LSD are like night and day. Heroin is night, a time to sleep and with sleep nothing comes. But with LSD it is like dawn, a new awakening, it expands your mind, it gives you a brand-new outlook on life."[10]

## DEPENDENCE, MDMA AND KETAMINE

However, it is possible to become dependent on ketamine or on MDMA, though much less likely than becoming dependent on, for example, cocaine or heroin. It's not certain, though, how addictive they are; estimates only exist for illicit use, which for ketamine is at very much higher doses than for psychiatric use.

Experts have concluded that you are as likely to become addicted to ketamine as you are to alcohol, and two-thirds as likely to become addicted to MDMA.[11]

A survey of nearly 4,000 people who use cocaine, MDMA and/or ketamine asked them about their dependence symptoms.

It tells a slightly different story that, out of cocaine, MDMA and ketamine users, the latter were the least likely to be dependent, at 15 percent.[12]

The addiction potential of MDMA and ketamine does make these two drugs somewhat paradoxical because both can also be used to treat addiction (see Chapter 7).[13, 14]

## HOW ADDICTIVE IS KETAMINE?

Ketamine dependence can lead to withdrawal symptoms, although these are more psychological than physical, such as anxiety and sleeplessness. Taken regularly, tolerance to ketamine develops quite fast; within weeks people have to take much bigger doses to get the same result.[15]

However, taking ketamine in a clinical setting eliminates the risk of addiction, because the patient only has a small and set number of treatments. Recreational doses also tend to be much higher than therapeutic doses, putting up the risk. That said, in the U.S. there have been reports of people abusing prescription ketamine.[16] There are some potential strategies that could prevent this: a national registry and prescription monitoring programs.[17, 18]

## HOW ADDICTIVE IS MDMA?

The question of how addictive MDMA is, like its effects on the brain, has been the subject of much debate. As you'd expect with a drug that most users take at festivals, parties and clubs, addiction to MDMA is largely a myth; generally, MDMA users only take it weekly or fortnightly.

Animal studies show that dependence on MDMA, as with ketamine, is less physical and more psychological and behavioral than dependence on strongly addictive drugs such as cocaine.[19] This is probably down to its strong effects on the serotonin system, which reduces the addictiveness caused by the effects on the dopamine system.

It's very rare that people seek treatment for MDMA dependence.[20] Some do stop using due to mental health concerns, but people tend to stop mainly because they move on in life, stop clubbing, or think the quality of the drug has dropped.[21]

It may be that MDMA doesn't lead to the usual kind of dependence. People do build up some tolerance but they don't keep escalating their use. There's a lack of a persistent desire to use, as people don't want to take it outside social contexts. They take MDMA for different motives than most other drugs: for its empathic, social and bonding effects.[22]

A 2010 survey showed that a quarter of MDMA users had three or more *DSM-IV* dependence symptoms, more than users of cocaine or ketamine. However, they were less likely to be concerned about this, want to use less, want help or have a persistent desire to use. Young women were the most likely to report dependence symptoms.[23]

There is good news when MDMA is used as a treatment: in our alcoholism treatment study we monitored for patients' further use after the two MDMA treatments we gave, and found people did not use it again, nor report any craving for it.[24]

# DO PSYCHEDELICS MAKE YOU MENTALLY ILL?

Among the many urban legends that persist about LSD is one that says anyone who takes it more than seven times is automatically classed as legally insane.[25]

This may be because a psychedelic trip is the nearest that any nonpsychotic person gets to being psychotic. The crossover in symptoms includes feelings of unreality, loss of control, ego dissolution (the loss of sense of self), inability to think straight, and sensory changes, although they are auditory in schizophrenia as opposed to visual in psychedelics.

In the 1950s and 1960s, there was a trend in American psychiatry to diagnose schizophrenia very widely for a range of conditions that we'd now put under the umbrella of personality problems or adjustment disorders.[26] Even so, there was no noticeable increase in schizophrenia diagnoses. This suggests there was no corresponding jump in people with psychosis due to the jump in the number of people taking LSD.

And if psychedelics did make you mentally ill, considering the 40,000-plus test subjects and millions of other users who took them in the 1950s and 1960s, you'd expect to see huge number of "acid casualties" around still. Two large-scale studies have compared the mental health of people in the U.S. population who've taken psychedelics with those who haven't.

One, by Norwegian researchers, used data from more than 130,000 people, just over 13 percent of whom had used psychedelics (LSD, psilocybin or mescaline). The researchers found those people were no more likely than those who hadn't to have had recent anxiety, depression, suicidality or any other serious psychological distress, or to have had mental health treatments. The researchers wrote: "Rather, in several cases psychedelic use was associated with lower rates of mental health problems."[27]

A second study looked at a sample of over 190,000 people. It found that having used psychedelics was associated with a reduced likelihood of psychological distress in the past month, and of suicidality in the past year.[28] Professor Matthew Johnson at Johns Hopkins University says: "We are not claiming that no individuals have ever been harmed by psychedelics. Anecdotes about acid casualties can be very powerful—but these instances are rare."[29]

One of the first study's authors, researcher Teri Krebs, says, "Psychedelics are psychologically intense, and many people will blame anything that happens for the rest of their lives on a psychedelic experience."[30]

## PSYCHEDELICS AND PSYCHOSIS

When researchers began to test LSD in the 1950s, the first theory was that it might help treat schizophrenia or psychosis. However, trials showed that the opposite was true: in most cases, it tended to make symptoms worse, in fact "markedly aggravated" them.[31]

Because psychedelic symptoms are so similar to some of those seen in psychosis, at that time researchers in the UK and the U.S. called psychedelics "psychotomimetics," that is they mimic psychosis. Researchers used them to try to work out what might be going on in disorders like schizophrenia.

In fact, psychedelics are still used in research to model the brain during psychosis. At Imperial College with Professor

Mitul Mehta at Kings College, we have used the psychedelic state as a shortcut for testing out and triaging possible new drugs for psychosis before trialing them in people with psychosis, a much more complicated thing to do. We've given healthy volunteers psilocybin to produce psychotic-like experiences, then drugs with antipsychotic potential. We found that one drug, saracatinib, originally designed as an anticancer agent but dropped because of its effects on the brain, reduced the effects of psilocybin in healthy volunteers. We are now moving to a trial in people with Parkinson's disease and psychosis.[32, 33]

Sometimes, some people who take psychedelics have a psychotic episode for a few days, then recover. The view now is that psychedelics don't *cause* lasting psychosis but can bring on lasting psychosis in people who have a propensity for it.

However, in an example of research coming full circle there's now work going on in Israel to see if microdosing— small and regular doses—might be useful for psychosis. The theory is that because psychedelics worsen psychosis through stimulating the serotonin 2A receptor, microdosing might have the opposite effect, i.e., desensitize this serotonin receptor, and so be antipsychotic.[34]

## THE IMPORTANCE OF SET AND SETTING

During the Cold War in the 1950s and early sixties, the CIA's secret testing program MK-ULTRA tested drugs including LSD in various universities, hospitals and research foundations as well as in prisons and the army. One idea was that LSD might be able to reprogram communists to become capitalists. They also tested LSD as a truth serum to be used during interrogations, and to see if it could be used as a weapon to confound the enemy's troops.

A lot of the data was destroyed at the beginning of the 1970s but there is enough to show that what was going on went against all ethical codes: for example, giving LSD to soldiers and civilians—including minors—without consent, explanation or warning,[35] and in horrific circumstances like being restrained and/or blindfolded, which had terrible consequences for the subjects' mental health.

A U.S. Senate hearing in 1975–1976 stated that this program led to the deaths of at least two Americans and "other participants in the testing programs may still suffer from the residual effects." It also said: "The nature of the tests, their scale and the fact that they were continued for years after the danger of surreptitious administration of LSD to unwitting individuals was known, demonstrate a fundamental disregard for the value of human life."[36, 37]

It's a scary example of the psychiatric profession being seduced into doing unethical testing on patients.

## TRUE OR FALSE: THE STRANGEST LSD STORIES

LSD is the drug with the most outlandish claims made about it, due to government propaganda from the 1960s and a scare campaign in the media as well as urban myths.

* LSD causes abortions and damage to sperm and unborn babies. **Untrue.**

A series of 1960s studies claimed to show this and, although it was disproved in 1971, the story has lived on. Also see the *National Examiner* headline from 1968: Girl Gives Birth to a Frog: Doctors Blame LSD.[38]

* LSD causes cancer. **Untrue.**

Another late-'60s scare story, based on research and case reports from patients. A recent population study found no association between psychedelic use and cancer.[39, 40]

* "Bad" acid gives you a bad trip. **Untrue.**

People taking psychedelics do often have challenging experiences, aka bad trips, including fear, anxiety and paranoia. Festivalgoers at Woodstock in 1969 were warned against the "brown acid" for this reason. It could be that "bad" acid is in fact another, less pleasant psychedelic, but it's more likely that a bad trip is caused by the wrong set and setting, or taking too much. Or being told you've just taken "bad" acid.

* LSD makes you think you can fly. **Possible.**

Deaths by misadventure are rare but do happen. There are reports of people jumping from buildings, for example, a young man who jumped from a second-floor balcony after taking mushrooms.[41]

Accidents do happen too, as in the high-profile and tragic death of teenager Arthur Cave, the son of the musician Nick Cave and the designer Susie Cave, who fell off a cliff near Brighton. The coroner, Veronica Hamilton-Deeley, said: "It [LSD] was taken by lads who were inquiring and experimenting and it's what kids do all the time, and most of the time...they get away with it, except on sad occasions like this." The lesson is, do not take a psychedelic if you're somewhere that could be dangerous.[42]

* LSD can make you suicidal. **True (very rarely).**
There have been rare case reports of suicide attempts and suicide after taking psychedelics; these seem to be linked either to unsafe settings or preexisting conditions.[43] But population studies suggest a drop in suicidality in people who have taken psychedelics, particularly psilocybin.[44]

Recently, reports made a lot of the fact that there were some subjects in a trial who had suicidal thoughts after being treated with psilocybin for treatment-resistant depression. But, in fact, the three people who this refers to were all nonresponders to the treatment.[45]

## HOW LIKELY IS A BAD TRIP?

We still don't know why some people, sometimes, have a bad trip. We know set and setting are key, which is why we have particular conditions and provide support from therapists during trials. We also know people with, for example, depression are more likely to have a difficult, challenging and painful experience. Having seen the kind of trips that our depressed patients went through, I wouldn't recommend that anyone with a mental health condition self-medicate with psychedelics.

## WILL LSD GIVE YOU FLASHBACKS?

Flashbacks, when you relive visuals and other drug symptoms long after the trip has ended, are a sort of never-ending mental hangover. However, they are a lot less common, shorter and milder than most people think. The theory used to be that LSD can hang around in the body for years, but we know now that it's eliminated by the body in hours rather than days.

There is a diagnosis, Hallucinogen Persisting Perceptual Disorder, if people have audio-visual and perceptual distortions for months or years. One study showed that HPPD sufferers had abnormalities in EEG readings, which suggests their brains were working differently.[46] However, HPPD is rare. In fact, out of the hundreds of patients—many with psychiatric problems—who have taken part in modern trials of psilocybin, there have been no serious adverse events, defined as "a reaction that results in death, is life-threatening, results in prolonged hospitalization or persistent or significant disability." There have been no reports of lasting psychosis or HPPD in modern trials of psilocybin, ayahuasca or LSD.[47]

## WHAT ARE THE HARMS OF KETAMINE?

Like MDMA, ketamine is most often used illicitly in clubs and at festivals, and people often mix it with other drugs.[48] Potential negative effects that come on quickly include confusion, disorientation, paranoia and agitation. As it's a dissociative anesthetic, if the person takes too much they can put themselves in danger by losing the sense of where they are and/or the ability to move or can become agitated. They can be vulnerable to crime and accident, even accidental death. People have frozen to death and drowned in rivers and in the bath. Twenty-one-year-old Louise Cattell drowned in the bath after taking ketamine. Her mother, Vicky Unwin, now campaigns for the legalization of drugs in order to allow people to make informed choices and to make them safer.[49, 50]

If ketamine is taken regularly, users build up tolerance quickly, which means having to take much bigger doses to get the same result. Heavy use commonly causes "K-cramps," painful spasms in the bladder. Over time, this leads to bladder and kidney inflammation, which can cause pain passing urine, needing to go more often, incontinence and, eventually, non-reversible bladder damage, ulcerative cystitis. At this stage, the only treatment is to remove the bladder, leaving the patient with a catheter for life. And because this puts you at risk of kidney and other infections, it takes years off your life expectancy.

Taking high doses of ketamine can also lead to brain damage, which is visible on brain scans, including lower volume of gray matter.[51] Symptoms are similar to dementia or to the negative state seen in schizophrenia. People become apathetic, have cognitive and motivational impairments, memory and attention problems, and become unable to plan. It's not clear how much of this damage is reversible, but it does appear to be dose dependent.[52]

Heavy users can also develop cognitive problems and delusional symptoms and are more likely to be depressed. Recreational users can also develop delusional symptoms. But ex-users don't appear to have any of these.[53]

## WHO IS EXCLUDED FROM PSYCHEDELIC TRIALS?

* Anyone diagnosed with psychosis or with a first-degree relative (parent or sibling) who has had psychotic episodes or schizophrenia.

  Reason: If someone is vulnerable to psychosis, there is a possibility a psychedelic might initiate an episode or make their altered perceptions and thinking processes more extreme.

* Anyone with a bipolar disorder.

  Reason: Possibility of provoking a manic episode.

* Anyone with a heart condition.

  Reason: These drugs can change heart rate and blood pressure.

The vast majority of harms from all these substances are linked to them being illicit, namely poisoning from adulteration or a too-high dose or being taken by people for whom they are not suitable, or in the wrong set or setting.

In the previous chapter, I described how you could get a clear idea of the real harm of taking ecstasy as a recreational drug by comparing it to riding. As these drugs become medicines, perhaps driving a car is a better analogy than riding a horse: without lessons and guidance, driving is dangerous. But with a driving instructor showing the novice how to do it and how to stay safe, it is not dangerous. The same is true of the prescription of therapy alongside MDMA, ketamine and psychedelics.

We do need more research on harm in the clinical setting and how to make these compounds safer. There are discussions under way looking at the way we report adverse effects in trials, in order to standardize them and make them comparable. And it would be very useful if psychedelics and MDMA could be taken out of Schedule 1 in order to make this research both cheaper and easier. The first country to do this was Australia, in February 2023, specifically psilocybin for treatment-resistant depression and MDMA for treatment-resistant PTSD.[54]

Although the previous two chapters have talked about a lot of possible harms, it is important to remember that most of these are based on people using these drugs in less-than-ideal ways.[55] And even then, for most people they don't lead to negative mental health consequences. In fact, if anything they tend to lead to positive ones.

# Conclusion

# WHAT ARE THE KEY QUESTIONS FOR THE FUTURE OF PSYCHEDELICS?

**JUST AS IN** the 1960s psychedelics transformed the shape of society, they may yet change the shape of medicine too. The combination of therapy and drugs is already a radical breakaway from our existing mental health treatments.

Now, the legalization of psilocybin and MDMA in Australia, as of July 1, 2023, may transform how we show that medicines work too. Because as well as giving hope to patients who've been failed by current medicines, the Australian initiative is accelerating our knowledge in this area.

In Australia, psychiatrists—who are trained in both psychotherapy and drug treatments—are allowed to give psilocybin for treatment-resistant depression and MDMA for treatment-resistant PTSD. The therapy is modeled on published trials that you've read about in this book. But the important new element is the academic oversight group that has developed, which is monitoring the treatment program for patients. Every patient is going onto a register, with details of their previous treatment failures and how they do on the new treatment.

This is a real-world evidence (RWE) trial, which for this group of patients can be a better approach than a randomized controlled trial (RCT). An RCT only tells you if a treatment is better than a placebo in a group of patients. And RCTs are both ethically challenging and scientifically limited in treating resistant patients with psychedelics. For example, these patients have often failed on several different prior treatments, and at this point giving them a placebo—required by an RCT—would be unethical. And as you can't fully blind psychedelics trials because of their psychoactive effects, the patient will know they are on placebo, which can make them feel worse (the nocebo effect).

When you can't do experiments, you have to look at evidence. RWE uses each patient as their own control, taking into account their treatment history. The DrugScience treatment analysis group has developed statistical approaches to analyze this RWE, based on Bayesian analysis, a well-established way of predicting the future from experience, used in climate change and economics.

After enough patients, we will know the likelihood of a patient achieving clinically meaningful progress over a period of time, say three months. We have used this approach to explore the impact of medical cannabis in children with treatment-resistant epilepsy. We found, using data from just 21 patients, that there is a 96 percent chance of the next child with similar epilepsy responding.

Doctors use this process all the time when they decide to try different drugs. But it's time to recognize that it's also a valid way to decide which drugs work. The eminent establishment figure

# CONCLUSION

Professor Michael Rawlins—ex-head of the National Institute for Health and Care Excellence (NICE) and the UK Medicines and Healthcare Products Regulatory Agency (MHRA)—long argued for Bayesian and RWE approaches to regulatory decision-making.[1]

But there's another radical element to the Australian program. There, both drugs will be provided by a charity, Mind Medicine. The reason I strongly support this is that it will make the treatment available to patients at an affordable price. It is reprehensible that patients may be put at risk of suicide because they can't access psychedelics until a pharmaceutical company decides to get a license to sell a drug for profit in their country.

Every day, there are millions of people around the globe being traumatized in wars and natural disasters. It's likely that just from the 2023 Turkish/Syrian earthquake, hundreds of thousands of people will end up suffering from depression or have PTSD. But it will likely be decades at best, and probably never, until a pharmaceutical company will make psychedelic therapy available in this region at a price people can afford. Given the evidence we have detailed in this book, trained doctors should be able to quickly access and use psychedelics for these patients when other options have failed.

## BREAKING THE LAWS

In order for essential research to happen, we need psychedelics to be rescheduled, taken out of Schedule 1 of the 1971 UN Conventions on Psychotropic Drugs, which states that they have no medicinal value. This can start at the state or country level—as in Oregon and Australia. There's a campaign at the UK level too, being driven by charities such as DrugScience and the Conservative Drug Reform Group plus the University of Manchester.

But it's also crucial that this happens at the level of the World Health Organization (WHO) and the United Nations (UN), who dictate drug policy to most countries. The WHO eventually conceded that cannabis has medicinal properties, and they will concede that psychedelics do, in due course. But the UN has yet to approve medical cannabis and so will probably also try to block psychedelics as a medicine. The good news is that more and more governments are accepting that they can act despite the intransigence of the UN.

Looking to the future, it's almost certain that more governments might become progressive in drugs laws as Portugal, the Netherlands and Canada have already done by decriminalizing personal drug possession.

I'm very sympathetic to the legal stance taken by the Latin American countries, that if a plant or mushroom grows naturally in their country, then it shouldn't be illegal. That's why I would undo the ban on magic mushrooms in the UK. At the moment, we have the absurd situation that picking a mushroom and immediately eating it is not illegal but taking it home is illegal possession and giving it to someone else is considered supply (dealing), even if no money changes hands.

If a natural substance is going to be sold in shops, things are moving toward a regulated market. This is happening in the Netherlands and in many U.S. states, led by Oregon, where they are developing regulations around growing natural products for other people's use. Any substance on sale would need

# CONCLUSION

to be labeled with clear details of its potency and active content, so that people are informed of appropriate dosing. In the same way as with cannabis, there should also be thresholds for content levels of active ingredients.

And as we are getting the benefit of these substances, it's important to make sure the indigenous peoples of where the natural products—for example, ayahuasca and ibogaine—originate are properly recompensed (for more, see Chapter 6).

If the government was going to legalize synthetics, such as LSD and MDMA, they would need a safe way to sell them, with some kind of oversight, in order to avoid vulnerable people taking them. This would likely be via a licensed shop, probably with a smart card allowing controlled access.

The funding for the development of non-psychedelic plastogens by the U.S. government illustrates that they are hoping people will be able to access treatments without any of these legal issues, or a trip. As doctors, if we can treat people without the risk of a bad trip it would be a good thing. It will be interesting to see what happens in this area; there are both commercial and academic researchers working on it, so it's likely results will be emerging in a few years. But will these drugs give a good therapeutic effect without a trip? I have a suspicion that, because the power of the psychedelics seems to come in part from the mystical experience (see Chapter 9), they won't have as good an effect.

# BETTER RESULTS

In the next ten years, as we come to know more about what psychedelics do in the brain, and as we collect more RWE, the way we're using psychedelics in psychotherapy will be refined and evolve.

In this book we've talked about a lot of conditions, but there are undoubtedly yet more that psychedelics could treat. It seems likely that they will be tested for all kinds of internalizing conditions—that is, any condition where a person finds it difficult to change their thinking processes.

Most exciting is the possibility of helping people with dyspraxia and/or dyslexia. It seems possible that the classic psychedelics may help as there is an abundance of 2A receptors in the areas of the brain that are in charge of speech and language. It will also be interesting to see trials looking at how psychedelic-assisted therapy can help people with non-substance addictions or behavioral addictions such as gambling addiction, gaming addiction and even sex addiction, where it's hard to intervene using traditional approaches.

Another promising approach is group therapy, as is common in ayahuasca ceremonies and some mushroom retreats. Group therapy was very widely used alongside LSD in the 1960s. Currently the way we do psychedelic therapy, with two therapists, makes it very expensive; as this shouldn't just be a therapy for the rich, it's important to find ways to cut costs. Another way to do this would be by reducing the number of therapists from two to one. And Compass Pathways have been giving healthy volunteers psychedelic experiences in groups, a stepping-stone toward giving psychedelics to patients in groups.[2]

However, group work does present challenges, for example, the possibility of individuals mistreating one another. And it's important that group therapy shouldn't be mandatory.

We also need to find ways for people who've had successful treatment to maintain their results if they fade away with time. Especially with treatment-resistant depression, many patients slip back into depressed thinking patterns. The following are some ideas for maintaining results, but they would need to be tested.

1) Top-ups of the full treatment every few months if the symptoms come back. We don't know if this will work or not as it hasn't been tested, but it's likely to.

2) Patient support groups. Groups such as Psypan and Rosalind Watts's ACER group have been set up to provide support to patients who've found that their symptoms improved. Groups will allow patients to share their recovery skills with one another.

3) Microdosing. It's possible that taking a small amount of a psychedelic might maintain enough neuroplasticity to stop mood slipping back to its old, depressed state.

4) It's also possible that, after psychedelics have reset the brain, other medications might work where they didn't before, protecting the person from relapse. That might be, for example, SSRIs or mood stabilizers such as lithium.

5) The same is true for talking therapies. After treatment, when the person has a deeper knowledge of the issues that are at the root of their condition, perhaps established

treatments such as CBT and other psychotherapies will be more likely to work.

6) Mindfulness and meditation approaches are looking promising when combined with psychedelics for treatment (see Chapter 6), by making use of psychedelics' mind-opening effects. But they might be useful for keeping someone well too.

7) Finding ways to relive the therapeutic experience. This might be via music or other traditional approaches to finding a mystical experience, such as drumming.

8) Non-psychedelic neuroplastogens. Even if these don't work as a treatment to, for example, lift depression or shift dependence, it's possible they might work to maintain wellness.

9) The newest technique we're trialing at Imperial College is neurostimulation, which uses electrical pulses to non-invasively turn on or off parts of the brain. It may be that doing this can make low doses of psychedelics work better. Studies show that it seems to work on animals, but it hasn't yet been trialed in humans. Neurostimulation may also add to the neuroplastic effects of psychedelics. And it could also be used to switch off the amygdala (alarm center of the brain) after or even during trauma. One day, someone with a gambling addiction might be able to lie in a scanner, see the roulette ball spinning, get zapped by the electrical pulse and have their dependence turned off.

The resurrection of psychedelics as a therapy is the most exciting thing I have seen in my long medical career. I started training in 1969! The fact that it has come about in the face of massive opposition from bureaucrats and other policy enforc- ers tells us that many other enlightened researchers and clini- cians feel the same as me, and have put in the huge efforts required to make this progress. This book is the story of some of their work, and mine too. The parallel growth of the brain science of psychedelics has helped convince the public that this is a serious and important human endeavor with a sound scientific underpinning, and one that in due course will likely provide mental health and neurological benefits to someone in most families in the world.

Although Aldous Huxley wrote the quote below in the 1950s, I can't think of a more appropriate way of ending. I hope, having read this book, you also feel the same.

*I suspect that these drugs [psychedelics] are destined to play a role in human affairs at least as important as alcohol has hitherto, and incomparably more beneficial.*

# NOTES

## CHAPTER ONE

1 https://www.tga.gov.au/news/media-releases/change-classification
-psilocybin-and-mdma-enable-prescribing-authorised-psychiatrists.
2 https://www.psychiatrictimes.com/view/psychiatry-long-view.
3 https://www.goodreads.com/quotes/542554-taking-lsd-was-a
-profound-experience-one-of-the-most.
4 https://www.theatlantic.com/health/archive/2014/09/the-accidental
-discovery-of-lsd/379564.
5 https://wchh.onlinelibrary.wiley.com/doi/pdf/10.1002/pnp.94.
6 https://www.ncbi.nlm.nih.gov/pmc/articles/PMC6787540/.
7 https://www.nytimes.com/2019/06/03/obituaries/james-ketchum
-dead.html.
8 https://www.theguardian.com/tv-and-radio/2022/jun/09/dr-delir
ium-and-the-edgewood-experiments-documentary.
9 https://www.thetimes.co.uk/article/drug-session-showed-me-huge
-vision-of-god-reveals-paul-mccartney-2s9p8kg5m.
10 https://pubmed.ncbi.nlm.nih.gov/21305914/.
11 https://druglibrary.net/schaffer/lsd/valencic.htm.
12 https://akjournals.com/view/journals/2054/7/1/article-p48.xml.
13 https://www.nytimes.com/2005/01/30/magazine/dr-ecstasy.html.
14 Huxley, A. (1977). *The Doors of Perception*. London, UK:
Harpercollins.
15 https://maps.org/images/pdf/books/lsdmyproblemchild.pdf.

# CHAPTER TWO

1 https://www.frontiersin.org/articles/10.3389/fphar.2018.00172/full#B47.
2 https://www.drugscience.org.uk/drug-information/#1551446554226-ae27d089-48e9.
3 https://pubmed.ncbi.nlm.nih.gov/33059356.
4 https://www.imperial.ac.uk/news/187706/potent-psychedelic-dmt-mimics-near-death-experience.
5 https://pubmed.ncbi.nlm.nih.gov/28129538.
6 https://pubmed.ncbi.nlm.nih.gov/33059356.
7 https://www.ncbi.nlm.nih.gov/pmc/articles/PMC4855588.
8 https://digitalcommons.buffalostate.edu/cgi/viewcontent.cgi?article=1034&context=exposition.
9 https://www.organism.earth/library/document/turn-on-tune-in-drop-out.
10 https://www.ncbi.nlm.nih.gov/pmc/articles/PMC3181823.
11 https://www.thelancet.com/pdfs/journals/lancet/PIIS0140-6736(08)60943-5.pdf.
12 https://www.ncbi.nlm.nih.gov/pmc/articles/PMC3181823.
13 https://www.inverse.com/article/14503-bicycle-day-albert-hofmann-lsd-acid-trip.
14 Hofmann, A. (1992). *LSD, My Problem Child: Reflections on Sacred Drugs, Mysticism, and Science*. Mt. View, CA: Wiretap.
15 https://www.ncbi.nlm.nih.gov/pmc/articles/PMC4813425.
16 https://www.wholecelium.com/blog/the-forgotten-story-of-valentina-wasson.
17 https://timeline.com/with-the-help-of-a-bank-executive-this-mexican-medicine-woman-hipped-america-to-magic-mushrooms-c41f866bbf37.
18 https://www.britishcouncil.org/voices-magazine/maria-sabina-one-of-mexicos-greatest-poets.
19 https://psychedelictimes.com/psychedelic-timeline.
20 https://www.drugscience.org.uk/drug-information/dmt.
21 https://www.thetimes.co.uk/article/santo-daime-the-drug-fuelled-religion-3mkf2pcfc58.
22 https://www.bbc.co.uk/programmes/p0438553.
23 https://www.mirror.co.uk/tv/tv-news/bbc-presenter-takes-drug-brazil-8693692.
24 https://psychedelictimes.com/the-universal-archetypes-of-ayahuasca-dreams-and-making-sense-of-your-own-visions/.
25 https://www.youtube.com/watch?v=Z9RFpR5VHN4.

26 https://www.ncbi.nlm.nih.gov/pmc/articles/PMC6378413.
27 https://smallpharma.com/press-releases/dmt-trials-phase-i-consistent-quality-psychedelic-response.
28 https://www.forbes.com/sites/natanponieman/2021/07/28/nasdaq-listed-mindmed-launches-human-trials-into-dmt-the-active-ingredient-in-ayahuasca/?sh=3067c9716c06.
29 https://pubmed.ncbi.nlm.nih.gov/20942780/.
30 https://thethirdwave.co/colorado-river-toad/.
31 https://www.youtube.com/watch?v=jYcnaYEzX7Y.
32 https://www.newyorker.com/magazine/2022/03/28/the-pied-piper-of-psychedelic-toads.
33 https://www.nytimes.com/2022/03/20/us/toad-venom-psychedelic.html?login=email&auth=login-email&login=email&auth=login-email.
34 https://pubmed.ncbi.nlm.nih.gov/30822141.
35 https://pubmed.ncbi.nlm.nih.gov/30982127.
36 Jay, M. (2021). *Mescaline: A Global History of the First Psychedelic*. London, UK: Yale University Press.
37 https://www.sciencedirect.com/science/article/abs/pii/S0028390822003537.
38 https://digitalcommons.law.seattleu.edu/ailj/vol9/iss1/6.
39 https://www.nature.com/articles/d41586-019-01571-2.
40 https://digitalcommons.law.seattleu.edu/ailj/vol9/iss1/6.
41 Boon, M. (2002). *The Road of Excess: A History of Writers on Drugs*. Cambridge, London: Harvard University Press.
42 https://www.bps.org.uk/psychologist/eye-fiction-heavenly-and-hellish-writers-hallucinogens.
43 https://michaelpollan.com/books/this-is-your-mind-on-plants.
44 https://pubmed.ncbi.nlm.nih.gov/36252614.
45 https://pubs.acs.org/doi/10.1021/acsptsci.1c00018.
46 https://pubmed.ncbi.nlm.nih.gov/32174803.
47 https://www.thetimes.co.uk/article/a-grieving-mothers-bittersweet-moment-of-triumph-over-legal-highs-nfq0xqvwtf7.
48 https://pubmed.ncbi.nlm.nih.gov/23994452.

# CHAPTER THREE

1 https://www.ncbi.nlm.nih.gov/pmc/articles/PMC5126726.
2 https://www.theguardian.com/commentisfree/2015/feb/18/ketamine-common-sense-britain-war-on-drugs.

3  https://www.jneurosci.org/content/35/33/11694.

4  https://www.nature.com/articles/nrd2172.pdf?origin=ppub.

5  Shulgin, A. T. (1992). *Controlled Substances: A Chemical and Legal Guide to the Federal Drug Laws*. Berkeley, CA: Ronin Pub.

6  https://doubleblindmag.com/inside-sasha-shulgins-lab/.

7  https://www.nytimes.com/2005/01/30/magazine/dr-ecstasy.html.

8  https://www.nytimes.com/2005/01/30/magazine/dr-ecstasy.html.

9  Shulgin, A. T., and Shulgin, A. (2021). *PiHKAL: A Chemical Love Story*. Berkeley, CA: Transform Press.

10 https://www.theguardian.com/science/shortcuts/2014/jun/03/alexander-shulgin-man-did-not-invent-ecstasy-dead.

11 https://blogs.scientificamerican.com/cross-check/talking-death-with-the-late-psychedelic-chemist-sasha-shulgin.

12 https://www.theguardian.com/science/from-the-archive-blog/2014/jun/03/shulgin-alexander-drugs-ecstasy-mdma.

13 Shulgin, A. T., and Shulgin, A. (2017). *TiHKAL: The Continuation*. Berkeley, CA: Transform Press.

14 https://www.bloodhorse.com/horse-racing/articles/235769/scopolamine-substance-in-middle-of-justify-scandal.

15 http://www.chm.bris.ac.uk/motm/scopolamine/scopolamineh.htm.

16 https://www.pbs.org/wnet/secrets/david – foran/204.

17 https://onlinelibrary.wiley.com/doi/10.1111/j.1556-4029.2010.01532.x.

18 https://www.vice.com/en/article/bvzdkw/tiktok-smelled-devils-breath-flower-hallucinogen-scopolamine.

19 https://www.newsweek.com/women-smell-poisonous-flower-angels-trumpet-worlds-scariest-drug-tiktok-1604733.

20 https://nationalpost.com/opinion/oliver-sacks-on-hallucinogenic-drugs-one-pill-makes-you-larger.

21 https://www.criver.com/products-services/discovery-services/pharmacology-studies/neuroscience-models-assays/alzheimers-disease-studies/scopolamine-induced-amnesia-model?region= 3696.

22 https://journals.lww.com/clinicalneuropharm/Citation/2014/03000/The_Potential_Role_of_Scopolamine_as.10.aspx.

23 https://www.reuters.com/article/astrazeneca-targacept-idUSL6E7M80SI20111108.

24 https://bwitilife.com/bwiti-tradition.

25 https://maps.org/research-archive/html_bak/ibogaine.html.

26 https://www.theguardian.com/society/2017/dec/10/ibogaine-heroin-addiction-treatment-gabon-withdrawal-danger-death.

27 https://maps.org/news/bulletin/in-memoriam-howard-lotsof.

28 https://www.herbalgram.org/resources/herbalgram/issues/87/table-of-contents/article3572/.

29 https://www.tandfonline.com/doi/full/10.1080/17425255.2021.1944 099.

30 https://www.nbcnews.com/id/wbna23573004.

31 https://www.nature.com/articles/s41598-020-73216-8.

32 https://thethirdwave.co/psychedelics/amanita-muscaria.

33 https://www.samwoolfe.com/2013/04/the-sacred-mushroom-and cross-by-john.html.

34 https://scfh.ru/en/papers/alcohol-and-hallucinogens-in-the-life-of-siberian-aborigines9124.

# CHAPTER FOUR

1 https://pubmed.ncbi.nlm.nih.gov/23616706.

2 https://www.ncbi.nlm.nih.gov/pmc/articles/PMC3277566/.

3 https://www.pnas.org/doi/10.1073/pnas.1518377113.

4 https://www.pnas.org/doi/10.1073/pnas.2218949120.

5 https://pubmed.ncbi.nlm.nih.gov/21609772.

6 https://pubmed.ncbi.nlm.nih.gov/17188554.

7 https://pubmed.ncbi.nlm.nih.gov/19580387.

8 https://www.ncbi.nlm.nih.gov/pmc/articles/PMC2268790.

9 https://royalsocietypublishing.org/doi/10.1098/rsif.2014.0873.

10 https://journals.sagepub.com/doi/full/10.1177/2050324520942345.

# CHAPTER FIVE

1 https://doi.org/10.1176/ajp.2006.163.11.1905.

2 https://www.ncbi.nlm.nih.gov/pmc/articles/PMC3050654.

3 https://jamanetwork.com/journals/jamapsychiatry/fullarticle/210962.

4 D. Nutt personal communication.

5 https://www.sciencedirect.com/science/article/pii/S089662730500156X.

6 https://pubmed.ncbi.nlm.nih.gov/20855296.

7 https://www.thelancet.com/journals/lanpsy/article/PII S2215-0366(16)30065-7.

8 https://www.ncbi.nlm.nih.gov/pmc/articles/PMC5367557.

9 https://pubmed.ncbi.nlm.nih.gov/27909164.

10 https://compasspathways.com/our-work/treatment-resistant.

# NOTES

11 https://compasspathways.com/wp-content/uploads/2022/05/Compass_APA_Poster_v5.1.pdf.
12 https://www.nejm.org/doi/full/10.1056/NEJMoa2032994.
13 https://www.ncbi.nlm.nih.gov/pmc/articles/PMC6442683.
14 https://www.ncbi.nlm.nih.gov/pmc/articles/PMC5410405.
15 https://www.ncbi.nlm.nih.gov/pmc/articles/PMC6082376.
16 https://news.unchealthcare.org/2020/06/roth-leads-26-9-million-project-to-create-better-psychiatric-medications.
17 https://pubmed.ncbi.nlm.nih.gov/35072760.
18 https://www.ncbi.nlm.nih.gov/pmc/articles/PMC6378413.
19 https://www.nature.com/articles/s41598-019-51974-4.
20 www.pnas.org/doi/10.1073/pnas.2218949120.
21 https://smallpharma.com/press-releases/dmt-trials-phase-i-consistent-quality-psychedelic-response.
22 https://pubs.acs.org/doi/10.1021/acsmedchemlett.1c00546.
23 https://www.biologicalpsychiatryjournal.com/article/S0006-3223(22)01553-0/fulltext.
24 https://tim.blog/2022/09/30/dr-john-krystal-ketamine.
25 https://www.cell.com/neuron/pdf/S0896-6273(19)30114-X.pdf.
26 https://www.nature.com/articles/srep46421.
27 https://pubmed.ncbi.nlm.nih.gov/21907221/.
28 https://ejnmmiphys.springeropen.com/articles/10.1186/s40658-021-00384-5.
29 https://www.oxfordhealth.nhs.uk/ketamine/.
30 Dr. Ben Sessa personal communication.
31 https://www.bbc.co.uk/news/health-54014522.
32 https://accp1.onlinelibrary.wiley.com/doi/abs/10.1002/jcph.1573.
33 https://www.oxfordhealth.nhs.uk/ketamine.
34 https://www.nejm.org/doi/full/10.1056/NEJMoa2032994.
35 https://jamanetwork.com/journals/jamapsychiatry/fullarticle/277263.

# CHAPTER SIX

1 Lee, M. A., and Shlain, B. (2001). *Acid Dreams: The Complete Social History of LSD: The CIA, the Sixties, and Beyond*. London: Sidgwick & Jackson.
2 Lee, M. A., and Shlain, B. (2001). *Acid Dreams: The Complete Social History of LSD: The CIA, the Sixties, and Beyond*. London: Sidgwick & Jackson.

# NOTES

3 Lee, M. A., and Shlain, B. (2001). *Acid Dreams: The Complete Social History of LSD: The CIA, the Sixties, and Beyond*. London: Sidgwick & Jackson.

4 https://pubmed.ncbi.nlm.nih.gov/9924845.

5 https://pubmed.ncbi.nlm.nih.gov/35107059.

6 https://journals.sagepub.com/doi/full/10.1177/2050324516683325.

7 https://journals.sagepub.com/doi/10.1177/2050324516683325.

8 Lee, M. A., and Shlain, B. (2001). *Acid Dreams: The Complete Social History of LSD: The CIA, the Sixties, and Beyond*. London: Sidgwick & Jackson.

9 Lee, M. A., and Shlain, B. (2001). *Acid Dreams: The Complete Social History of LSD: The CIA, the Sixties, and Beyond*. London: Sidgwick & Jackson.

10 https://www.cambridge.org/core/journals/the-psychiatrist/article/dr-ronald-arthur-sandison/350EA7A27FA66EB827A66C22ACDB73E6.

11 https://pubmed.ncbi.nlm.nih.gov/34326773.

12 https://pubmed.ncbi.nlm.nih.gov/28447622.

13 https://pro.psycom.net/psychedelic-assisted-therapy/how-to-administer-psychedelic-assisted-therapy.

14 https://psychiatryinstitute.com/podcast/about-psychedelics-pioneers-richards.

15 https://www.frontiersin.org/articles/10.3389/fpsyt.2021.586682/full

16 https://www.drrosalindwatts.com.

17 https://www.drugscience.org.uk/podcast/59-how-to-become-a-psychedelic-therapist-with-dr-rosalind-watts.

18 https://www.sciencedirect.com/science/article/abs/pii/S0924977X16300165.

19 https://www.ncbi.nlm.nih.gov/pmc/articles/PMC5893695.

20 https://psycnet.apa.org/record/1973-07379-001.

21 https://www.ncbi.nlm.nih.gov/pmc/articles/PMC5893695.

22 https://pubmed.ncbi.nlm.nih.gov/22114193.

23 https://pubmed.ncbi.nlm.nih.gov/29020861.

24 https://pubmed.ncbi.nlm.nih.gov/30965131.

25 https://www.sciencedirect.com/science/article/pii/S0149763422002135.

26 https://cris.maastrichtuniversity.nl/en/publications/depression-mindfulness-and-psilocybin-possible-complementary-eff e.

27 https://www.frontiersin.org/articles/10.3389/fpsyg.2022.866018/full.

28 https://pubmed.ncbi.nlm.nih.gov/33732156/.

29 https://akjournals.com/view/journals/2054/5/2/article-p57.xml.

30 https://pubmed.ncbi.nlm.nih.gov/28631526.

31 https://www.iceers.org/about-us.

32 https://pubmed.ncbi.nlm.nih.gov/25242255.
33 https://maps.org/2019/05/24/statement-public-announcement-of
-ethical-violation-by-former-maps-sponsored-investigators.
34 https://psychedelicspotlight.com/maps-sex-elder-abuse-scan
dals-health-canada-reviews-mdma-trials.
35 https://pubmed.ncbi.nlm.nih.gov/35031907.
36 https://pubmed.ncbi.nlm.nih.gov/34267683.
37 https://pubmed.ncbi.nlm.nih.gov/35924888.
38 https://www.frontiersin.org/articles/10.3389/fpsyt.2022.831092/.
39 https://maps.org/2010/05/05/how-to-work-with-difficult
-psychedelic-experiences.

# CHAPTER SEVEN

1 https://www.thymindoman.com/aa-co-founder-bill-wilsons-first
-vision-account.
2 https://en.wikipedia.org/wiki/Charles_B._Towns.
3 Cheever, S. (2005). *My Name Is Bill: Bill Wilson: His Life and the Creation of Alcoholics Anonymous*. New York: Washington Square.
4 https://silkworth.net/alcoholics-anonymous/ebby-t-the-man
-who-carried-the-message-to-bill-w.
5 https://www.gov.uk/government/publications/review-of-drugs-
phase-one-report/review-of-drugs-summary.
6 https://www.gov.uk/government/publications/delivering-better-oral
-health-an-evidence-based-toolkit-for-prevention/chapter-12-alcohol#
fnref:3:1.
7 https://www.gov.uk/government/publications/2010-to-2015
-government-policy-cancer-research-and-treatment/2010-to-2015
-government-policy-cancer-research-and-treatment.
8 https://ukhsa.blog.gov.uk/2019/02/14/health-matters-preventing
-cardiovascular-disease.
9 https://www.recoveryspeakers.com/if-there-is-a-god-will-he-show
-himself-bill-w-recounts-his-spiritual-experience.
10 https://www.ncbi.nlm.nih.gov/books/NBK99377.
11 Kennedy, quoted in Lee, M. A., and Shlain, B. (2001). *Acid Dreams: The Complete Social History of LSD: The CIA, the Sixties, and Beyond*. London: Sidgwick & Jackson.
12 https://www.ncbi.nlm.nih.gov/books/NBK99377.
13 https://jamanetwork.com/journals/jamapsychiatry/article
-abstract/490914.

14  https://www.who.int/health-topics/tobacco#tab=tab_1.
15  https://pubmed.ncbi.nlm.nih.gov/25213996.
16  https://www.ncbi.nlm.nih.gov/pmc/articles/PMC5641975.
17  https://clinicaltrials.gov/ct2/show/results/NCT01943994.
18  https://clinicaltrials.gov/ct2/show/results/NCT01943994.
19  Adapted from Baler and Volkow 2006.
20  Adapted from Baler and Volkow 2006.
21  https://onlinelibrary.wiley.com/doi/full/10.1111/adb.12980.
22  https://pubmed.ncbi.nlm.nih.gov/9250944.
23  https://psychology.exeter.ac.uk/kare/.
24  https://ajp.psychiatryonline.org/doi/10.1176/appi.ajp.2021
    .21030277.
25  https://pubmed.ncbi.nlm.nih.gov/25586396.
26  https://jamanetwork.com/journals/jamapsychiatry/fullarticle/
    2795625.
27  https://www.nytimes.com/2022/08/25/health/psilocybin-mushrooms
    -alcohol-addiction.html.
28  https://journals.sagepub.com/doi/full/10.1177/0269881121991792.
29  https://www.heraldopenaccess.us/openaccess/how-well-are
    -patients-doing-post-alcohol-detox-in-bristol-results-from-the
    -outcomes-study.
30  https://pubmed.ncbi.nlm.nih.gov/28402682.
31  https://pubmed.ncbi.nlm.nih.gov/27870477.
32  https://ajp.psychiatryonline.org/doi/10.1176/appi.ajp.2019
    .18101123.

# CHAPTER EIGHT

1  https://transformdrugs.org/blog/mdma-history-and-lessons
   -learned-part-1#_ftn3.
2  https://transformdrugs.org/blog/mdma-history-and-lessons-learned
   -part-1#_ftn3.
3  https://www.tga.gov.au/news/media-releases/change-classification
   -psilocybin-and-mdma-enable-prescribing-authorised-psychiatrists.
4  https://time.com/6253702/psychedelics-psilocybin-mdma
   -legalization/.
5  https://www.kcl.ac.uk/news/pioneering-trial-of-mdma-treatment
   -hopes-to-help-veterans-with-ptsd.
6  https://cks.nice.org.uk/topics/post-traumatic-stress-disorder/back
   ground-information/prevalence.

7  https://www.ptsd.va.gov/understand/common/common_veterans
   .asp.

8  https://watson.brown.edu/costsofwar/files/cow/imce/papers/2021/
   Suitt_Suicides_Costs%20of%20War_June%2021%202021
   .pdf.

9  https://maps.org/mdma/ptsd/mapp1.

10 https://www.nature.com/articles/s41591-021-01336-3.

11 https://maps.org/news/media/press-release-fda-grants-breakthrough
   -therapy-designation-for-mdma-assisted-psychotherapy-for-ptsd
   -agrees-on-special-protocol-assessment-for-phase-3-trials.

12 https://www.ncbi.nlm.nih.gov/pmc/articles/PMC4578244.

13 https://www.gov.uk/government/statistics/drug-misuse-findings
   -from-the-2012-to-2013-csew/drug-misuse-findings-from-the-2012
   to-2013-crime-survey-for-england-and-wales.

14 https://journals.sagepub.com/doi/full/10.1177/2050324518767442.

15 https://maps.org/wp-content/uploads/2015/10/MDMAIB13th
   EditionFinal22MAR2021.pdf.

16 https://journals.sagepub.com/doi/full/10.1177/2050324518767442.

17 https://journals.sagepub.com/doi/full/10.1177/2050324518767442.

18 https://journals.sagepub.com/doi/full/10.1177/2050324518767442.

19 https://journals.sagepub.com/doi/full/10.1177/2050324518767442.

20 https://journals.sagepub.com/doi/pdf/10.1177/2050324518767442.

21 https://www.thetimes.co.uk/article/ann-shulgin-obituary
   -6qdcfhrcb.

22 https://podcasts.apple.com/au/podcast/37-maps-with-rick-doblin/
   id1474603382?i=1000523137520.

23 https://journals.sagepub.com/doi/10.1177/2050324515578080.

24 https://journals.sagepub.com/doi/10.1177/2050324515578080.

25 https://journals.sagepub.com/doi/10.1177/2050324515578080.

26 https://chriskresser.com/treating-ptsd-with-mdma-dr-michael
   -mithoefer.

27 https://www.heroicheartsproject.org.

28 https://www.heroicheartsuk.com.

29 https://compasspathways.com/our-work/post-traumatic-stress
   -disorder-ptsd.

30 https://ajp.psychiatryonline.org/doi/10.1176/appi.ajp.2020
   .20121677.

31 https://chriskresser.com/treating-ptsd-with-mdma-dr-michael
   -mithoefer.

32 https://pubmed.ncbi.nlm.nih.gov/29728331.

33 https://www.ncbi.nlm.nih.gov/pmc/articles/PMC3122379.

34 https://maps.org/2014/01/27/a-manual-for-mdma-assisted-therapy
   -in-the-treatment-of-ptsd.

35  https://www.drugscience.org.uk/podcast/41-mdma-and-couples
    -counselling/.
36  https://maps.org/news-letters/v23n1/v23n1_p10-14.pdf.
37  https://www.nature.com/articles/s41591-021-01336-3.
38  https://www.ncbi.nlm.nih.gov/pmc/articles/PMC3573678.
39  https://maps.org/news-letters/v23n1/v23n1_p10-14.pdf.
40  https://pubmed.ncbi.nlm.nih.gov/24495461.
41  https://pubmed.ncbi.nlm.nih.gov/24495461.
42  https://pubmed.ncbi.nlm.nih.gov/24495461.
43  https://www.nature.com/articles/npp201735.
44  https://pubmed.ncbi.nlm.nih.gov/24345398.
45  https://www.nature.com/articles/s41591-021-01385-8.
46  https://pubmed.ncbi.nlm.nih.gov/24345398.
47  https://pubmed.ncbi.nlm.nih.gov/28858536.
48  https://pubmed.ncbi.nlm.nih.gov/27592327.
49  https://pubmed.ncbi.nlm.nih.gov/26408071.
50  https://www.nature.com/articles/s41586-019-1075-9.
51  https://www.nature.com/articles/s41598-020-75706-1.
52  https://www.frontiersin.org/articles/10.3389/fpsyt.2022.944849/full.
53  https://pubmed.ncbi.nlm.nih.gov/30196397.
54  https://maps.org/2022/06/29/mdma-assisted-group-therapy
    -for-ptsd-among-veterans-study-will-proceed-following-successful
    -safety-negotiations.
55  https://clinicaltrials.gov/ct2/show/NCT03752918.
56  https://maps.org/mdma.
57  https://pubmed.ncbi.nlm.nih.gov/33601929/.
58  https://scienceofparkinsons.com/tag/lawrence.
59  https://www.bbc.co.uk/science/horizon/2000/ecstasyagony.shtml.
60  https://pubmed.ncbi.nlm.nih.gov/22345403.

# CHAPTER NINE

1  https://lettersofnote.com/2010/03/25/the-most-beautiful-death.
2  https://pubmed.ncbi.nlm.nih.gov/26029972.
3  https://pubmed.ncbi.nlm.nih.gov/11324179/.
4  https://www.ncbi.nlm.nih.gov/pmc/articles/PMC1896303.
5  https://pubmed.ncbi.nlm.nih.gov/34958455/.
6  https://pubmed.ncbi.nlm.nih.gov/20819978/.
7  https://www.ncbi.nlm.nih.gov/pmc/articles/PMC4086777/.
8  https://pubmed.ncbi.nlm.nih.gov/27909164/.

9 https://www.ncbi.nlm.nih.gov/pmc/articles/PMC5367557/#b ibr16-0269881116675513.

10 https://www.thelancet.com/journals/eclinm/article/PIIS2589 -5370(20)30282-0/fulltext.

11 https://psychedelicpress.co.uk/products/modes-of-sentience.

12 https://www.frontiersin.org/articles/10.3389/fpsyg.2023.1128589/ full.

13 https://www.frontiersin.org/articles/10.3389/fnhum.2016.00269/ full.

14 https://www.frontiersin.org/articles/10.3389/fphar.2017.00974/full.

15 https://journals.sagepub.com/doi/abs/10.1177/0022167817709585.

16 https://quoteinvestigator.com/2019/06/20/spiritual.

17 https://www.vice.com/en/article/yvqqpj/psilocybin-the-mush room-and-terence-mckenna-439.

18 https://www.theguardian.com/science/2011/apr/06/martin-rees -templeton-prize.

19 Strassman, R. (2001). *DMT: The Spirit Molecule: A Doctor's Revolutionary Research into the Biology of Near-Death and Mystical Experiences*. Rochester, VT: Park Street Press.

20 https://pubmed.ncbi.nlm.nih.gov/31745107/.

21 https://www.pnas.org/doi/10.1073/pnas.2218949120.

22 https://pubmed.ncbi.nlm.nih.gov/23881860.

23 https://www.nature.com/articles/s41598-019-45812-w.

24 https://journals.sagepub.com/doi/abs/10.1177/0269881117736919.

25 https://pubmed.ncbi.nlm.nih.gov/30174629.

26 https://www.ncbi.nlm.nih.gov/pmc/articles/PMC6220878/.

27 https://pubmed.ncbi.nlm.nih.gov/35939083/.

28 https://www.youtube.com/watch?v=_92ZqEmeOFs.

29 https://www.frontiersin.org/articles/10.3389/fnhum.2016.00269/ Full.

30 https://pubmed.ncbi.nlm.nih.gov/24508051.

31 Shulgin, A. (2021). *PiHKAL: A Chemical Love Story*. Berkeley, CA: Transform Press.

32 Pollan, M. (2018). *How to Change Your Mind*. London, UK: Penguin Press.

33 https://www.atpweb.org/jtparchive/trps-23-91-01-001.pdf.

34 https://clinicaltrials.gov/ct2/show/NCT02421263.

35 https://www.lucid.news/pioneering-clergy-of-diverse-religions -embrace-psychedelics.

# CHAPTER TEN

1 https://www.ncbi.nlm.nih.gov/pmc/articles/PMC8965278.
2 https://www.nature.com/articles/s41386-022-01301-9.
3 https://www.biologicalpsychiatryjournal.com/action/showPdf?pii=S0006-3223%2822%2901553-0.
4 https://www.theguardian.com/science/2022/mar/08/demand-grows-for-uk-ministers-to-reclassify-psilocybin-for-medical-research.
5 https://pubmed.ncbi.nlm.nih.gov/16801660.
6 https://clusterbusters.org/resource/psilocybin-and-lsd-treatment/.
7 https://www.medrxiv.org/content/10.1101/2022.07.10.22277414v1.
8 https://journals.sagepub.com/doi/abs/10.1177/0333102410363490.
9 https://pubmed.ncbi.nlm.nih.gov/25903763.
10 https://clusterbusters.org/resource/lsa-seeds-of-the-vine.
11 https://pubmed.ncbi.nlm.nih.gov/29722608.
12 https://tryptherapeutics.com/assets/documents/pdf/Tryp-Therapeutics_Pain-and-Psychedelics_Castellanos_2021-10-02-221023_huug.pdf.
13 https://www.nature.com/articles/d41586-022-02878-3.
14 https://pubmed.ncbi.nlm.nih.gov/32067950.
15 https://algernonpharmaceuticals.com/dmt-stroke-program.
16 https://www.ncbi.nlm.nih.gov/pmc/articles/PMC7472664.
17 https://link.springer.com/article/10.1007/s00213-019-05417-7.
18 https://www.eatingdisorderhope.com/information/anorexia/anorexia-death-rate.
19 https://proto.life/2022/02/psychedelics-offer-new-route-to-recovery-from-eating-disorders/.
20 https://www.frontiersin.org/articles/10.3389/fpsyt.2021.735523/full.
21 https://www.lidsen.com/journals/neurobiology/neurobiology-05-02-102.
22 https://clinicaltrials.gov/ct2/show/NCT04052568.
23 https://www.biologicalpsychiatryjournal.com/article/S0006-3223(22)00773-9/fulltext.
24 https://www.nature.com/articles/s41591-023-02455-9.
25 https://compasspathways.com/compass-pathways-launches-phase-2-clinical-trial-of-psilocybin-therapy-in-anorexia-nervosa/.
26 https://link.springer.com/article/10.1007/s40519-020-01000-8.
27 https://pubmed.ncbi.nlm.nih.gov/30474794/.
28 https://tryptherapeutics.com/updates/tryp-therapeutics-announces-interim-results-for-its-phase-ii-clinical-trial-for-the-treatment-of-binge-eating-disorder-with-psilocybin-assisted-psychotherapy.

29  https://www.mdpi.com/2076-3425/12/3/382.
30  https://pubmed.ncbi.nlm.nih.gov/17196053.
31  https://clinicaltrials.gov/ct2/show/NCT03300947.
32  https://icpr2020.net/speakers/francisco – moreno.
33  https://www.youtube.com/watch?v=2JP3dtERUS0.
34  https://www.youtube.com/watch?v=cIuU68pNVow.
35  https://www.orchardocd.org/participate-in-research/the-psilocd-study-recruitment-is-now-open-now-closed/.
36  https://www.irishnews.com/magazine/entertainment/2022/02/10/news/rory-bremner-says-adhd-is-my-best-friend-and-my-worst-enemy-2585610.
37  https://pubmed.ncbi.nlm.nih.gov/30925850.
38  https://cks.nice.org.uk/topics/attention-deficit-hyperactivity-disorder/background-information/prevalence.
39  https://www.ncbi.nlm.nih.gov/pmc/articles/PMC6753862.
40  https://www.sciencedirect.com/science/article/pii/S0924977X20309111.
41  https://pubmed.ncbi.nlm.nih.gov/28576350.
42  https://discovery.ucl.ac.uk/id/eprint/1513566/1/Journal%20of%20Attention%20Disorders-2016-Mowlem-1087054716651927.pdf.
43  https://clinicaltrials.gov/ct2/show/NCT05200936.

# CHAPTER ELEVEN

1  https://www.itv.com/news/2022-11-09/the-tiktokers-who-claim-microdosing-psychedelic-drugs-helps-their-mental-health.
2  Fadiman, J. (2011). *The Psychedelic Explorer's Guide: Safe, Therapeutic, and Sacred Journeys.* Rochester, VT: Park Street Press.
3  https://www.drugscience.org.uk/podcast/56-microdosing-with-james-fadiman/.
4  https://pubmed.ncbi.nlm.nih.gov/30925850.
5  Waldman, A. (2017). *A Really Good Day: How Microdosing Made a Mega Difference in My Mood, My Marriage, and My Life.* Deckle Edge.
6  https://microdosinginstitute.com/microdosing-101/benefits.
7  https://pubmed.ncbi.nlm.nih.gov/30685771.
8  https://pubmed.ncbi.nlm.nih.gov/31853557.
9  https://www.cambridge.org/core/journals/the-psychiatrist/article/

dr-ronald-arthur-sandison/350EA7A27FA66EB827A66C22AC
DB73E6.

10  https://pubmed.ncbi.nlm.nih.gov/27210031.

11  https://pubmed.ncbi.nlm.nih.gov/33059356.

12  https://academic.oup.com/brain/article-abstract/145/9/2967/
6648879.

13  https://journals.sagepub.com/doi/full/10.1177/2045125320950567.

14  https://hal.archives-ouvertes.fr/hal-02491823/document.

15  https://journals.sagepub.com/doi/full/10.1177/0269881119857204.

16  https://pubmed.ncbi.nlm.nih.gov/30726251.

17  https://www.nature.com/articles/s41598-022-14512-3.

18  https://www.nature.com/articles/s41598-021-81446-7.

19  Hall, K. T. (2022). *Placebos*. Boston, MA: The MIT Press.

20  https://www.nytimes.com/2018/11/07/magazine/placebo-effect
-medicine.html.

21  https://www.theguardian.com/science/2022/oct/08/placebos-expert
kathryn-t-hall-eff ect-painkillers-interview.

22  https://psyarxiv.com/cjfb6.

23  https://onlinelibrary.wiley.com/doi/abs/10.1111/adb.13143.

24  https://www.nature.com/articles/s41598-021-81446-7.

25  https://www.frontiersin.org/articles/10.3389/fphar.2018.00897/full.

26  https://www.nejm.org/doi/full/10.1056/nejmoa2032994.

27  https://elifesciences.org/articles/62878.

28  https://pubmed.ncbi.nlm.nih.gov/33082016.

29  https://www.nature.com/articles/s41598-021-01811-4.

30  https://lighthouse.mq.edu.au/article/april-2022/clinical-trial-to
-test-psychedelics-in-treating-depression.

31  https://pubmed.ncbi.nlm.nih.gov/16149323.

32  https://pubmed.ncbi.nlm.nih.gov/16149324.

33  https://pubmed.ncbi.nlm.nih.gov/16149326.

# CHAPTER TWELVE

1  https://publications.parliament.uk/pa/cm200506/cmselect/
cmsctech/1031/1031.pdf.

2  https://www.thelancet.com/journals/lancet/article/PIIS0140-6736
(07)60464-4/fulltext.

3  https://assets.publishing.service.gov.uk/government/uploads/
system/uploads/attachment_data/file/119088/mdma-report.pdf.

4  https://journals.sagepub.com/doi/abs/10.1177/0269881108099672.

# NOTES

5  https://publications.parliament.uk/pa/cm201213/cmselect/cmhaff/184/120619.htm.

6  https://www.channel4.com/news/articles/uk/jacqui%2Bsmith%2Bin%2Bexpenses%2Brow/2932757.html.

7  https://www.theguardian.com/politics/2009/mar/29/jacqui-smith-expenses-film.

8  https://www.crimeandjustice.org.uk/sites/crimeandjustice.org.uk/files/Estimating%20drug%20harms.pdf.

9  https://www.theguardian.com/politics/2009/nov/02/drug-policy-alan-johnson-nutt.

10  https://www.thelancet.com/journals/lancet/article/PIIS0140-6736(10)61462-6/fulltext.

11  https://www.nature.com/articles/srep08126?message1.

12  https://pubmed.ncbi.nlm.nih.gov/29482434/.

13  https://www.research.lancs.ac.uk/portal/en/publications/mdmapowder-pills-and-crystal-the-persistance-of-ecstasy-and-the-poverty-of-policy(b9c5030b-cb26-4125-9418-33d996770f65).html.

14  https://www.ons.gov.uk/peoplepopulationandcommunity/birthsdeathsandmarriages/deaths/bulletins/deathsrelatedtodrugpoisoninginenglandandwales/2021registrations.

15  https://www.justice.gov/archive/ndic/pubs2/2580/odd.htm.

16  https://www.ussc.gov/sites/default/files/pdf/news/congressional-testimony-and-reports/drug-topics/200105_RtC_MDMA_Drug_Offenses.pdf.

17  https://www.sciencedirect.com/science/article/pii/S0140673698043293.

18  https://pubmed.ncbi.nlm.nih.gov/12351788.

19  https://www.webmd.com/mental-health/news/20020926/one-night-of-ecstasy-can-damage-brain.

20  https://www.ncbi.nlm.nih.gov/pmc/articles/PMC194116.

21  https://pubmed.ncbi.nlm.nih.gov/10867554/.

22  https://www.aclu.org/press-releases/court-rejects-harsh-federal-drug-sentencing-guideline-scientifically-unjustified.

23  https://www.aclu.org/sites/default/files/field_document/mccarthy_decision.pdf.

24  https://www.courthousenews.com/ecstasy-has-same-legal-penalties-as-cocaine.

25  https://www.ncbi.nlm.nih.gov/pmc/articles/PMC1705495.

26  https://pubmed.ncbi.nlm.nih.gov/19195429.

27  https://jamanetwork.com/journals/jamapsychiatry/fullarticle/1151061.

28  https://www.nature.com/articles/npp2011332.

29 https://jamanetwork.com/journals/jamapsychiatry/fullarticle/211305.
30 https://journals.sagepub.com/doi/abs/10.1177/0269881118767646.
31 https://www.ncbi.nlm.nih.gov/pmc/articles/PMC1705495.
32 https://pubmed.ncbi.nlm.nih.gov/21646575.
33 https://pubmed.ncbi.nlm.nih.gov/34894842/.
34 https://pubmed.ncbi.nlm.nih.gov/19585107.
35 https://pubmed.ncbi.nlm.nih.gov/30579220.
36 https://link.springer.com/article/10.1007/s00213-007-0837-5.
37 https://pubmed.ncbi.nlm.nih.gov/17548754.
38 https://www.drugscience.org.uk/mdma – research.
39 https://www.sciencedirect.com/science/article/abs/pii/S0955395919303445?via%3Dihub.
40 https://www.theguardian.com/society/2015/mar/22/doctors-urged-to-talk-openly-with-patients-about-drug-taking-for pleasure.
41 https://transformdrugs.org/blog/mdma-history-and-lessons-learned-part-2.
42 https://volteface.me/post-pandemic-trends-in-the-uk-mdma-market.
43 https://www.dazeddigital.com/artsandculture/article/31093/1/why-are-pills-so-strong-right-now.
44 https://www.sciencedirect.com/science/article/abs/pii/S0955395919303445.
45 https://volteface.me/post-pandemic-trends-in-the-uk-mdma-market.
46 https://www.drugscience.org.uk/drug-checking-the-evidence-is-building.
47 https://www.sciencedirect.com/science/article/abs/pii/S0955395921001675.
48 https://pubmed.ncbi.nlm.nih.gov/14673568/.
49 https://www.drugscience.org.uk/drug-information/mdma.

# CHAPTER THIRTEEN

1 https://www.vice.com/en/article/5g984z/why-are-psychedelics-illegal-368.
2 https://digitalcommons.buffalostate.edu/cgi/viewcontent.cgi?article=1034&context=exposition.
3 https://www.unodc.org/pdf/india/publications/DAIIM_Manual_TTK/3-17.pdf.

# NOTES

4 https://www.ncbi.nlm.nih.gov/pmc/articles/PMC4813425/.
5 https://jamanetwork.com/journals/archneurpsyc/article-abstract/
652297.
6 https://www.nature.com/articles/1300711.
7 https://www.nature.com/articles/npp201786.
8 https://pubmed.ncbi.nlm.nih.gov/21842159.
9 https://www.tandfonline.com/doi/abs/10.3109/0095299930
9001618.
10 https://jamanetwork.com/journals/jamapsychiatry/article-abstract/
490914.
11 https://www.thelancet.com/journals/lancet/article/PIIS0140
-6736(10)61462-6/fulltext.
12 https://www.sciencedirect.com/science/article/abs/pii/S095539
5914001728.
13 https://journals.sagepub.com/doi/abs/10.1177/0269881121991792.
14 https://discovery.ucl.ac.uk/id/eprint/10044586/1/Lawn%20
Ketamine_NP_revised_references_WL_5thDec_clean.pdf.
15 https://medwinpublishers.com/ACT/ACT16000106.pdf.
16 https://www.nytimes.com/2023/02/20/us/ketamine-telemedicine
.html.
17 https://www.thelancet.com/journals/lanpsy/article/PIIS2215
-0366(17)30102-5/fulltext.
18 https://www.drugrehab.com/2016/05/20/bill-expands-prescrip
tion-drug-monitoring-programs.
19 https://pubmed.ncbi.nlm.nih.gov/19836170.
20 https://karger.com/nps/article/60/3-4/137/233556/Can-the
-Severity-of-Dependence-Scale-Be-Usefully.
21 https://pubmed.ncbi.nlm.nih.gov/14533132.
22 https://pubmed.ncbi.nlm.nih.gov/36162335.
23 https://www.sciencedirect.com/science/article/abs/pii/S0955
395914001728.
24 https://www.biosciencetoday.co.uk/results-of-mdma-treatment-trial
-for-alcohol-use-disorder-published.
25 https://en.wikipedia.org/wiki/Urban legends about drugs # Babysitter
places baby_in_the_oven_while_high_on_LSD.
26 https://jamanetwork.com/journals/jamapsychiatry/article-abstract
/490493.
27 https://pubmed.ncbi.nlm.nih.gov/23976938.
28 https://pubmed.ncbi.nlm.nih.gov/25586402/.
29 https://www.bioedonline.org/news/nature-news-archive/no
-link-found-between-psychedelics-psychosis.
30 https://www.bioedonline.org/news/nature-news-archive/no
-link-found-between-psychedelics-psychosis.

31  https://www.semanticscholar.org/paper/Effects-of-mescaline–and
    -lysergic-acid-(d-LSD-25).-Hoch-Cattell/1fa4f9ea1842d8a80738a
    77406927eccaba0acf1.

32  https://clinicaltrials.gov/ct2/show/NCT03661125.

33  https://gtr.ukri.org/projects?ref=MR%2FR005931%2F1.

34  https://pubmed.ncbi.nlm.nih.gov/36280752.

35  https://journals.sagepub.com/doi/pdf/10.1177/.

36  https://www.intelligence.senate.gov/sites/default/files/
    hearings/95mkultra.pdf.

37  https://www.smithsonianmag.com/smart-news/what-we-know
    -about-cias-midcentury-mind-control-project-180962836.

38  https://www.sciencenews.org/article/1967-lsd-was-briefly
    -labeled-breaker-chromosomes.

39  https://journals.sagepub.com/doi/abs/10.1177/02698811221117536.

40  https://www.psypost.org/2022/10/large-national-survey-suggests
    -that-the-use-of-psychedelics-is-not-associated-with-lifetime
    -cancer-development-64070.

41  https://pubmed.ncbi.nlm.nih.gov/30548541/.

42  https://www.theguardian.com/music/2015/nov/10/nick-caves-son
    -died-from-fall-after-taking-lsd-inquest-hears.

43  https://www.psychiatrist.com/read-pdf/39225/.

44  https://www.nature.com/articles/s41598-022-25658-5.

45  https://www.nejm.org/doi/full/10.1056/NEJMoa2206443.

46  https://pubmed.ncbi.nlm.nih.gov/8912957.

47  https://pubmed.ncbi.nlm.nih.gov/35107059.

48  https://assets.publishing.service.gov.uk/government/uploads/
    system/uploads/attachment_data/file/264677/ACMD_ketamine
    _report_dec13.pdf.

49  https://www.theguardian.com/society/2012/mar/03/louise-death
    -drugs.

50  https://transformdrugs.org/blog/welcoming-our-new-trustee.

51  https://www.frontiersin.org/articles/10.3389/fnana.2022.795231/
    full.

52  https://assets.publishing.service.gov.uk/government/uploads/
    system/uploads/attachment_data/file/119098/ketamine-report.pdf.

53  https://pubmed.ncbi.nlm.nih.gov/19133891.

54  https://www.tga.gov.au/resources/publication/scheduling-decisions
    -final/notice-final-decision-amend-or-not-amend-current-poisons
    -standard-june-2022-acms-38-psilocybine-and-mdma.

55  https://pubmed.ncbi.nlm.nih.gov/35107059.

# CONCLUSION

1 https://www.ncbi.nlm.nih.gov/pmc/articles/PMC4954394/pdf/579 .pdf.
2 https://compasspathways.com/our-work/comp360-psilocybin-in -healthy-participants.

# ACKNOWLEDGMENTS

**MANY PEOPLE HAVE** inspired and supported me in my career as a researcher and psychiatrist. Particular thanks are owed to Professor David Grahame-Smith, director of the MRC Clinical Pharmacology Unit in Oxford. He gave me my crucial first break in science by taking me into his unit as a clinical research fellow, and showed great patience as I explored the new world of neuropsychopharmacology. I am also very grateful to Professor Michael Gelder, the head of the psychiatry department in Oxford, for supporting my transition into clinical research, and the Wellcome Trust for funding me in this. Also to Dr. John Lewis, research director of Reckitt & Colman, for setting up the Bristol University psychopharmacology unit that allowed me the independence to develop my own research career.

The research on psychedelics and MDMA described in this book has only been made possible by the efforts of many friends and colleagues who have given up their time and intellect to help with the studies. These are Professor Val Curran, Dr. Ben Sessa, Dr. Tim Williams, Dr. Suresh Muthukumaraswamy, Dr. David Erritzoe, Dr. Mark Bolstridge, Professor Enzo Tagliazucchi, Dr. Rosalind Watts, Dr. James Rucker, Dr. Rick Strassman, Professor Mitul Mehta, George Goldsmith and Dr. Ekaterina Malievskaia, Dr. Meg Spriggs, Professor Celia

# ACKNOWLEDGMENTS

Morgan, Dr. Rick Doblin, Peter Hunt and Tania de Jong of Mind Medicine Australia, and many others, including the whole team at DrugScience.

Thanks to the Alexander Mosley Charitable Trust, the Saisei Foundation, Sanjay Singhal, Anton Bilton, the Charles Engelhard Foundation, Beckley Psytech, Orchard OCD and the many smaller donors who support our psychedelic research.

Special thanks are due to Amanda Feilding and the Beckley Foundation for their intellectual input and financial support over the past decade. Also thanks go to Channel 4 Science for funding the MDMA study, COMPASS for supporting the psychedelic work and the Mosley Foundation for supporting the second psilocybin depression study.

It's been a great pleasure to work again with Brigid Moss. To a large extent, this book succeeds because of her skill in structuring large amounts of information, her diligent analysis and questioning of the science and history of psychedelics and her clarity of writing.

Finally, and most importantly, I need to thank Professor Robin Carhart-Harris for his exceptional efforts in performing this series of challenging and controversial studies and then bringing them to fruition with such remarkable conclusions. Without his relentless endeavors and intellectual inputs, this new chapter in human brain research would not have been written.

# INDEX

# INDEX

# INDEX

# INDEX

# INDEX

# INDEX

# ABOUT THE AUTHOR

DAVID NUTT
MBBChir (Cambridge), DM (Oxford),
FRCP, FRCPsych, FBPhS
FMedSci, hon DLaws (Bath)

David Nutt is a psychiatrist and the Edmund J. Safra Professor of Neuropsychopharmacology in the Division of Brain Science, Department of Medicine, Hammersmith Hospital, Imperial College London. His research area is psychopharmacology—the study of the effects of drugs on the brain, from the perspectives of both how drug treatments in psychiatry and neurology work, and why people use and become addicted to some drugs such as alcohol. To study the effects of drugs in the brain he uses state-of-the-art techniques such as brain imaging with PET and fMRI plus EEG and MEG. This research output has led to over 500 original research papers that put him in the top 0.1 percent of researchers in the world. He has also published a similar number of reviews and book chapters, eight government reports on drugs and 37 books, including one for the general public, *Drugs Without the Hot Air*, that won the Transmission Prize in 2014.

He has held many leadership positions in science and medicine including presidencies of the European Brain Council, the British Association of Psychopharmacology, the British Neuroscience Association and the European College of Neuropsychopharmacology. He is currently Founding Chair of DrugScience.org.uk, a charity that researches and tells the truth about all drugs, legal and illegal, free from political or other interference. He also currently holds visiting professorships at the Open University and University of Maastricht.

David broadcasts widely to the general public both on radio and television and has a podcast (see below). In 2010 the *Times Eureka* science magazine voted him one of the 100 most important figures in British science, and he was the only psychiatrist on the list. In 2013 he was awarded the John Maddox Prize from Nature / Sense about Science for standing up for science and in 2017 a Doctor of Laws hon causa from the University of Bath.

en.wikipedia.org/wiki/David Nutt
www.sciencemag.org/content/343/6170/478.full
https://www.imperial.ac.uk/people/d.nutt
https://www.imperial.ac.uk/medicine/departments/
www.drugscience.org.uk/drug-science-podcast